# THE ECOLOGY OF TEMPORARY WATERS

# THE ECOLOGY OF TEMPORARY WATERS

D. DUDLEY WILLIAMS BSc, Dip. Ed MSc, PhD
Division of Life Sciences
Scarborough Campus
University of Toronto

CROOM HELM
London & Sydney

TIMBER PRESS
Portland, Oregon

Croom Helm Ltd, Provident House, Burrell Row,
Beckenham, Kent, BR3 1AT

Croom Helm Australia, 44-50 Waterloo Road,
North Ryde, 2113, New South Wales

British Library Cataloguing in Publication Data

Williams, D. Dudley
The ecology of temporary waters.
1. Aquatic ecology
I. Title
574.5'263 QH541.5.W3

ISBN 0-7099-5211-2

First published in the USA 1987 by
Timber Press,
9999S. W. Wilshire,
Portland, OR 97225,
USA



ISBN 0-88192-081-9

Printed and bound in Great Britain by
Biddles Ltd, Guildford and King's Lynn

# Contents

To the memory of my father,

Frank Williams,
1919 - 1986

# Preface

The primary role of this book is to introduce the reader to, and hopefully stimulate interest in, the ecology of temporary aquatic habitats. The book assumes that the reader will have, already, some general knowledge of ecology but this is not essential.

Temporary waters exhibit amplitudes in both physical and chemical parameters which are much greater than those found in most waterbodies. The organisms that live in these types of habitats have, therefore, to be very well adapted to these conditions if they are to survive. Survival depends largely on exceptional physiological tolerance or effective immigration and emigration abilities. Examples of such adaptations are given throughout the book and it is hoped that these will aid the reader in gaining an insight into the structure and function of plant and animal communities of these unusual habitats. The final chapter suggests field and laboratory projects that should be useful to students in school and university studies.

I thank Professor Ernest Naylor for providing facilities in the School of Animal Biology, U.C.N.W., Bangor, where part of the text was written. Dr. Henri Tachet, Departement de Biologie animale et Ecologie, Universite Claude Bernard, Lyon provided details of the fish ponds at Dombes; Dr. David Dudgeon, Department of Zoology, University of Hong Kong provided information on the temporary streams of that area; Dr. Richard Marchant, Museum of Victoria, Australia and Peter Outridge, University of Toronto provided information on the Magela Creek system, Northern Territories; Rama Chengalath, Invertebrate Zoology Division, National Museum, Ottawa provided data on the distribution of branchiopods in Canada. Dr. Ron Dengler, University of Toronto gave advice on matters botanical. Geraldine Dunn drew many of the figures and she and Mrs. Sybella Hendley typed the original manuscript. D. Harford, L. McGregor, T. Westbrook and A. Hollingsworth assisted with the illustrations. Mrs. K.A. Moore assisted with the index. I would like to thank also the many authors who gave me permission to reproduce tables and figures from their publications.

Finally, I would like to thank my wife, Dr. Nancy Williams for a critical reading of the manuscript and for excusing me from many domestic duties so that I could write.

D. Dudley Williams  Toronto, 1987

# 1 INTRODUCTION TO TEMPORARY WATERS

## 1.1 What are temporary waters?

> We have short time to stay, as you,
> We have as short a Spring,
> As quick a growth to meet decay,
> As you or any thing.
>
> \- Robert Herrick, 1591-1674

Temporary ponds and streams are found in many parts of the world. They are essentially natural bodies of water which experience a recurrent dry phase of varying duration. The emphasis is on the cyclical nature of drought in these habitats as permanent water bodies are also capable of going dry in exceptional years. In such cases, however, most of the permanent water fauna will be wiped out because it is not adapted to survive such conditions. Cyclical temporary water bodies, on the other hand, select for species which are adapted to these conditions.

## 1.2 Classification of temporary waters:

Classification of temporary aquatic habitats has been attempted by few people. One way of doing it is on the basis of size and Table 1.1 categorizes various types of temporary waters at three levels: micro-, meso- and macrohabitats. Perhaps the best criteria to use are length and intensity of the dry period as these relate more to the biology of these habitats.

Length of the dry phase is probably best divided simply into seasonal, annual and greater than annual - but cyclical. Intensity of the drought is important because, for example, two habitats which both remain dry for four months of the year might have different moisture-retaining capacities in their substrates allowing the survival of totally different faunas. As with all systems of classification, there are bound to be exceptions which do not fit any of the categories, for example, Lake Eyre in southern Australia only fills with water every half century or so (Mawson, 1950). Can this really be called cyclical? The majority of animals that would colonize such a lake would die when it dried up, while only highly specialized forms which are

Table 1.1: Classification of temporary water habitats based on size.

| | |
|---|---|
| Microhabitats: | axils of plant leaves, e.g. bromeliads, tree hollows, tin cans, broken bottles and other containers, foot prints, tyre tracks, empty shells |
| Mesohabitats: | flood-plain pools, temporary streams, temporary ponds, snow-melt pools, monsoon rain-pools |
| Macrohabitats: | large old river-beds – periodically flooded, shallow oxbow lakes, drying lakes, drying lakeshores, alpine lakes |

(adapted from Decksbach, N.K. von, 1929)

relatively few in number, would be able to span the fifty-six years between fill ups. Again, the occurrence of some temporary water bodies in Britain affords us an example of a misfit. *Triops* (a notostracan or tadpole shrimp), was first recorded in 1738 from a "temporary" pond, it was next recorded in 1837 and again in 1948, after a lapse of 111 years (Schmitt, 1971). Although it did not appear during all that time, it could be hatched from dried mud obtained from the pond. It was concluded that this species required a period of desiccation to stimulate hatching and Britain, not being renowned for its dry climate, was unsuitable for promoting the establishment of its habitats, that is cyclical temporary ponds. Nevertheless when the permanent ponds in which its eggs lay dried up, it hatched.

Klimowicz (1959) attempted to classify small ponds in Poland on the basis of their molluscan faunas. Granted, some snail and clam species are very resistant to a temporary loss of water, each having its own particular range of drought resistance, but some species have drought-resistant properties that differ between populations. Nevertheless, in some instances it is possible to speak of a "species X - temporary water", but unless this species has great dispersal and colonization powers, such a classification will have little meaning outside of a rather limited geographical area. Some other criteria for classifying temporary waters are given in later chapters which deal with specific environmental or biological topics.

## 1.3 Importance of temporary waters:

Temporary waters in general, and temporary ponds in particular, are extremely important from a scientific point of view as they are populated by quite a wide variety of species with interesting, if not unique, physiological and behavioural properties. In addition, because these species form part of functional communities, they make good subjects for the study of ecology, especially as the habtitats are small; yet the data resulting from these microcosms are applicable to larger situations and ecological theory in general.

In an applied sense, Mozley (1944) drew attention to the fact that temporary ponds are a neglected natural resource. When a pond dries up in early summer, in temperate regions, the bed becomes part of the terrestrial habitat. This habitat is well fertilized, due to the excrement left by the aquatic organisms, and thus supports a considerable biomass of land plants during the terrestrial phase. The terrestrial community, in turn, leaves a legacy of organic debris which can be used the following spring by the aquatic community. Mozley proposed that it should be possible to use temporary ponds for the rotational harvesting of stocked fish fry during the aquatic phase and a field crop such as oats during the terrestrial phase; each community would be nourished by the leavings of the other.

Temporary waters have deleterious aspects too. Many temporary waters, especially in the tropics, are the breeding places for the vectors of disease organisms. Intermittent ponds and ditches, irrigation canals, marshes and flooded areas support large numbers of aquatic snails and mosquitoes (Styczynska-Jurewicz, 1966). Snails are host to the blood trematode *Schistosoma*, a debilitating and eventually fatal parasite of humans and cattle, and the liver fluke *Fasciola hepatica*. Mosquitoes transmit malaria, yellow fever, dengue and viral encephalitis, while sucking the blood of man and domestic animal.

The importance of studying temporary waters has recently been underlined by W.D. Williams (1985) who states that ".....the extent of limnological references to temporary waters is not in accord with their widespread occurrence and abundance, ecological importance, nor limnological interest".

# 2 THE PHYSICAL FACTORS THAT GOVERN THE FORMATION OF TEMPORARY WATERS

## 2.1 The runoff cycle:

Figure 2.1 shows the basic components of the runoff cycle which are important in the formation of both permanent and temporary ponds and streams. Precipitation is the most important source of water so far as the cycle is concerned. Before this water even touches the ground surface it can be intercepted several times by trees and other vegetation; water trapped on these exposed surfaces is very quickly evaporated by wind. The water that reaches the soil surface is taken up by infiltration and the rate at which this occurs depends on the type of soil and its aggregation. At this stage, in some exceptional clayey soils, water may collect on the surface in small depressions and form puddles and even small trickles. Both tend to be short-lived, as the water they contain is usually absorbed quite quickly by soil cracks and patches of more permeable soil over which it may run. Such water bodies are referred to as ephemeral streams and are apparent only after periods of high storm intensity or snow melts; they seldom contain any animal or plant life.

Infiltrated water near the soil surface is subject to direct evaporation back into the atmosphere due to air currents and uptake and subsequent transpiration by surface vegetation such as grass. If the intensity of precipitation at the soil surface is greater than the infiltration capacity of the soil, and if all the puddles have been filled, then overland flow starts. When this reaches a stream channel it becomes surface runoff. If the topography is such that the water cannot flow away in a channel, then it collects in a low point and forms a pond.

Some of the water that penetrated the now-saturated soil will reach the stream channel or pond as interflow, usually where a relatively impervious layer is found close to the soil surface. The rest of the infiltrated water, that which has penetrated as far as the groundwater table, eventually will also reach the stream or pond as baseflow or groundwater flow.

In summary then, the water flowing in a stream or collecting in a pond is derived from the following sources: overland flow,

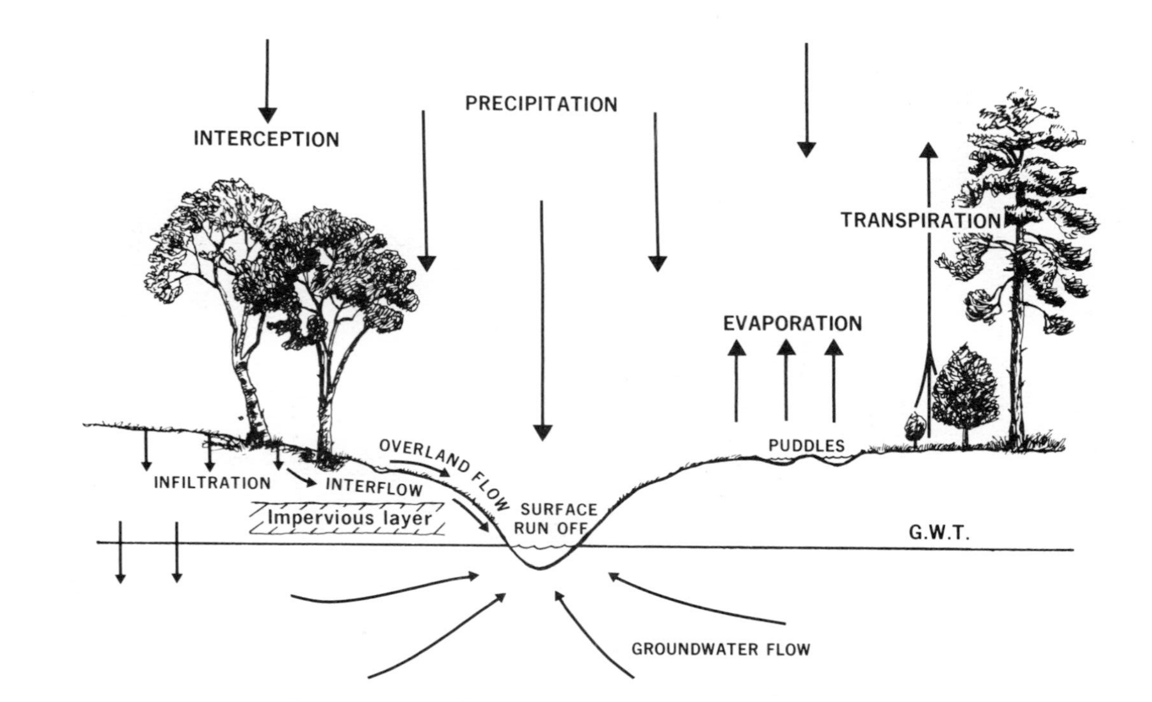


Figure 2.1: The basic components of the runoff cycle that contribute to the water in a pond or stream.

interflow, groundwater flow and direct precipitation on the water body itself.

Infiltration is perhaps the single most important factor in the regulation of temporary waters, for it determines how precipitation will be partitioned into the categories of overland, inter- and groundwater flow. Horton (1933) defined infiltration capacity as the maximum rate at which a given soil can absorb precipitation in a given condition. In the initial phase of infiltration the attraction of the water by capillary forces of the soil is of great importance, although the effect of these forces in medium to coarse-grained soils is only minor after the infiltration front has penetrated more than a few feet. These capillary forces are greatest within fine-grained soils which have low initial moisture (Davis and DeWeist, 1966). Air trapped between the soil particles may have an effect opposite to that of soil structure as, at first, the infiltration rate will be slowed down as the advancing front of infiltrating water will have a tendency to expel any air it meets and this may result in the formation of pockets of dry soil which will form barriers to water movement. However, as the front continues, some air may be dissolved and the rate of advance will speed up.

The condition of the soil is also of great importance as, for example, a bare soil surface will be directly exposed to rain which will tend to compact the soil and also wash small particles into open cracks and holes. This in turn will have the effect of reducing infiltration as the rain continues and may lead to overland flow. Conversely, a dense cover of vegetation will protect the soil surface so that compaction and the filling up of cracks will be less. Further, the roots of these plants may hold the soil open and increase the normal infiltration rate.

Taking the above factors into consideration, we could speculate that a large number of temporary streams and ponds would be supported by an area of land with a clay-loam soil under heavy cultivation, where many of the large stands of trees and much of the bush have been cleared, where bare soil is more common than pasture and where wire fences are preferred to earth-bank hedgerows (Figure 2.2). Were this land to be left uncultivated it probably would support a much smaller number of permanent streams and ponds instead of a large number of temporary ones.

Figure 2.2: Kirkland Creek, Ontario, Canada during late spring (upper photograph) and late summer (lower photograph).

## 2.2 Components of subsurface water:

We have seen what factors contribute to and influence the regime of water in a temporary pond or stream, but we have still to consider what happens when the water stops and the habitat becomes part of the terrestrial environment. To do this, it is perhaps best to begin by looking at the way in which subsurface waters are classified. Figure 2.3 shows the components of subsurface water beneath a dry pond or stream bed to consist of four zones known as soil water, intermediate water, capillary water and ground or phreatic water. There

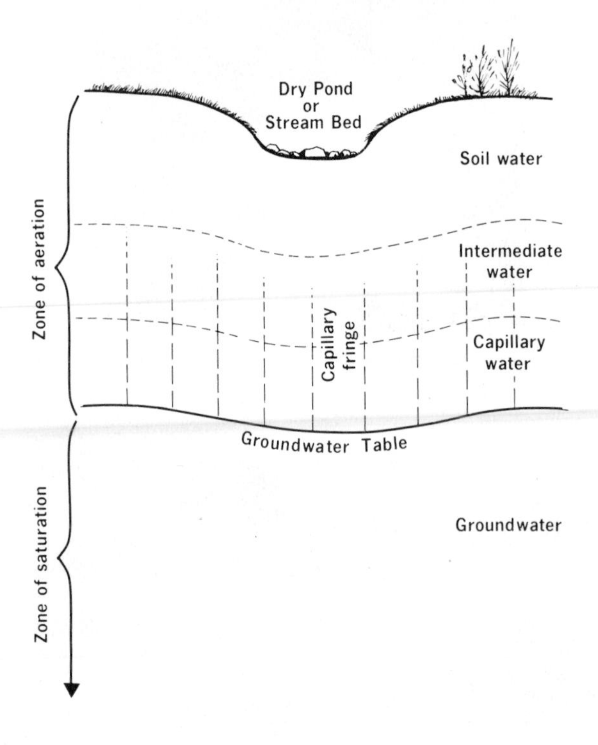


Figure 2.3: The components of subsurface water.

are other zones beneath that of the groundwater such as the internal water; these include water in unconnected pores and water in chemical combination with rocks, but since they are beyond the access of most aquatic organisms they need not concern us here.

Soil water is subject to large fluctuations in amount as a response to transpiration and direct evaporation, and it is this feature which separates it from the other unsaturated zones.

The intermediate water zone lies beneath that of the soil water and separates it from the saturated zone. The water here is sometimes referred to as suspended water since although it can move downwards in response to gravity, it also can move upwards into the soil water zone should this become very dry. This zone is variable in size, being greatest in arid regions and absent in moist areas.

The lower limit of the intermediate zone is continuous with the capillary fringe. This fringe is irregular in outline and consists of water moving up through the soil (by capillary action) from the lower parts of the capillary water zone which may be as fully saturated as the groundwater. In areas of fine-grained soils, where recharge is active, the capillary fringe may extend well into the intermediate zone.

The groundwater table (G.W.T.) separates the capillary fringe from the next water zone, the groundwater, where all the material is saturated. The groundwater table is commonly approximated to the level to which water will rise in a well and it can move up or down in response to recharge. When a stream is flowing, the G.W.T. will be near the level of the stream surface and, as it recedes, it will lower the level of the stream (or pond) unless it is offset by sudden precipitation and consequent overland flow and interflow. If it continues to recede, the stream will cease to flow and only a few pools will remain in the channel. Further recession will result in disconnection of the G.W.T. from the pools which will soon vanish due to evaporation. If the G.W.T. remains fairly close to the ground surface, the capillary fringe may extend up to the surface of the stream or pond bed and thus provide a moist environment for those aquatic animals that have sought refuge by burrowing. It is in situations such as this, where the G.W.T. is close to the soil surface and soil moisture is high that a relatively small input of moisture from rain, is sufficient to produce a substantial relaxation of moisture tension within the soil pores (Carson and Sutton, 1971). The result is that the G.W.T. may rise quickly and cause water to reappear in the pond basin. It is for this reason too that temporary streams may be observed to restart for short periods after quite small amounts of rainfall.

Any further drop in the G.W.T. will result in the setting up of the other subsurface water zones mentioned and hence subjecting animals to the large fluctuations of water content that we have seen are characteristic of the soil water zone. Deeper burrowing of active forms to reach the saturated zone, or the co-ordination of a special drought-resistant stage in the life cycle will now be necessary if these animals are to survive. In some cases, it is possible that a few species may obtain sufficient moisture from the condensation of dew on the ground surface to remain in the soil water zone.

## 2.3 Basin and channel formation:

### *2.3.1 Pond basins*

Pond basins may be formed by natural, geologic processes or by anthropogenic ones. Naturally-produced ponds frequently result from glacial activity, involving both erosional and depositional processes.

Ice scour of flat bedrock areas typically creates basins of varying size and depth. These ponds receive all their water either directly from precipitation or by surface runoff. Other types of erosion-formed lakes and ponds are cirques, which are amphitheatre-shaped basins carved at the heads of glaciated valleys, and paternoster lakes, which are chains of ponds formed in the bottoms of these valleys.

Different pond types result from the deposition of glacial debris. For example, retreating glaciers leave large deposits of ground moraine over wide areas. Where these overlie impermeable till, vast areas of "wetlands" and very shallow lakes and ponds have been created. Similarly, kettle ponds have been left in many areas by the melting of ice masses buried in the moraine. Deposits left at the bottoms of valleys may confine melting ice and so form basins.

Besides these erosionally- and depositionally-formed basins, glaciers have produced depressions through alternate freezing and thawing of the ground surface, resulting in subsidence. Shallow arctic and antarctic ponds are formed in this way (Reid, 1961). Where many of these ponds merge the result is a thermokarst lake (Rex, 1961).

Solution ponds are formed in regions where soluble rock has been dissolved by water. Infiltrating surface water freshly-charged with carbon dioxide (forming weak carbonic acid) is particularly effective as a solvent.

Small, deep ponds may result from plunge-pools at the base of dried-up waterfalls. Large, shallow ponds may be formed in "oxbow" fashion in any wide valley through which a river meanders, and shallow deltaic ponds are formed by sediment deposition at a river mouth.

Percolating waters may deposit an insoluble iron-pan layer in sandy, permeable soils and above this a pond may form. There are many examples of this type of pond in Surrey in England as well as in Sologne, France (Bowen, 1982).

Meteor impact is known to have created both large and small basins but these appear to be rare, or are perhaps rarely recognized as such.

Uprooting of trees by storms commonly creates small shallow basins that drain readily in sandy soils but which may contain water for several months in clay soils (Figure 2.4).

Figure 2.4: Small temporary pools created by uprooted trees.

Man-made basins result from industrial activities such as mining, quarrying, landscaping, etc. and also from ancient rural activities (especially in the Old World). Examples of these basins include: watering holes; peat-digging holes; moats (not just around castles, as, in the thirteenth century, moats were common features added to houses of gentry and farmers alike as status symbols [Taylor, 1972]); fish ponds (see Chapter 8); decoys (long, reed-lined, shallow channels leading from a lake and used for luring wild ducks into a trapnet); dewponds (shallow, nineteenth century, clay-lined ponds for collecting rainwater and runoff); armed ponds (watering holes with several spreading arms used for sharing water between several fields); saw pits (a practice begun in the fourteenth century, where one of two men sawing a log lengthwise stood in a pit beneath the log); and charcoal pits (an ancient process producing charcoal by burning wood buried in a pit) (Rackham, 1986).

In both ancient and modern times, ponds, ditches, trenches and craters have been created during warfare.

### 2.3.2 *How many ponds?*

As many temporary ponds are small they do not appear on most topographical maps. It is therefore difficult to estimate their abundance. By careful study of fine-scale Ordnance Survey maps and applying a correction factor for small ponds (<6m in diameter) not marked, Rackham (1986) estimated the total number of ponds in England and Wales to be 800,000 (or 5.4 ponds/$km^2$) around 1880. He considered this period to be when the total number of natural and man-made basins was at a maximum - this includes both basins that are permanently and intermittently filled. The frequency of occurrence of these basins was not uniform across the country (Figure 2.5), being least dense in mountainous areas (e.g. 0.12/$km^2$) and most dense in areas of ancient agriculture and ancient woodlands (e.g. 115/$km^2$). Undoubtedly, this analysis will have missed many of the very small temporary pond basins and so the final total for England and Wales may well be in excess of one million.

### 2.3.3 *Stream channels*

Most river valleys and stream channels are formed by erosion, so the processes of formation are much less diverse than for pond basins. Although some large rivers can dry up, intermittency is more characteristic of small headwater or tributary streams. Irrigation ditches represent man-made temporary streams and their frequency varies according to regional agricultural practice and climate.

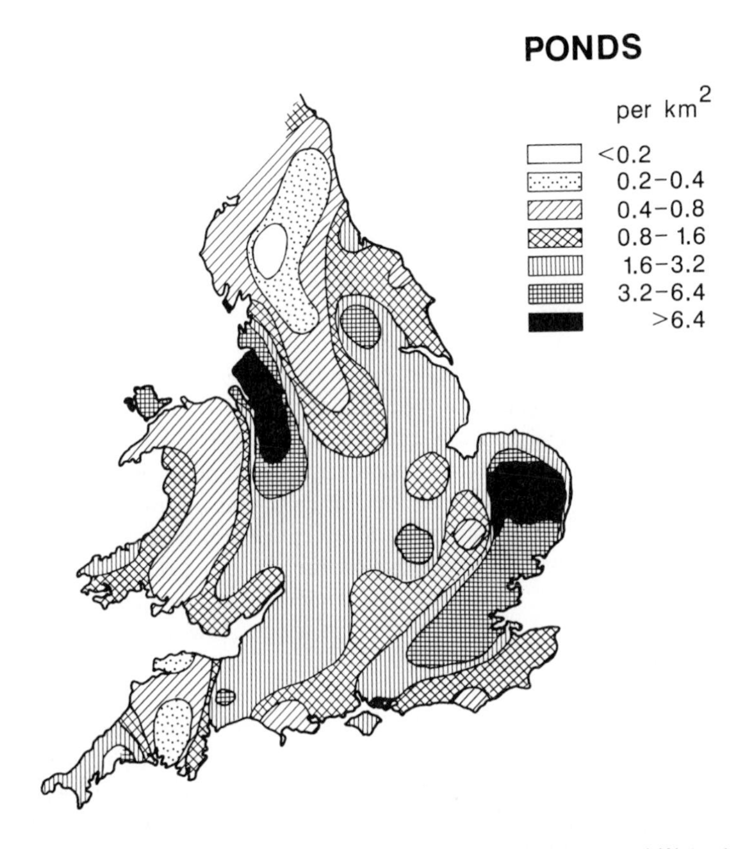


Figure 2.5: Distribution and density of ponds in England and Wales in the 1920 s (redrawn from Rackham, 1986).

### 2.3.4 *How many streams?*

Temporary streams are sometimes marked on fine-scale topographical maps as broken lines surrounded by symbols indicating a bog or marsh. Estimates of temporary stream densities for southern Ontario, Canada range between 0.04 and 1.10/km$^2$, with a mean of 0.7/km$^2$. This figure may be representative of temperate regions of the world having clayey soils under heavy cultivation (see Section 2.1).

# 3 ABIOTIC FEATURES OF TEMPORARY WATERS

## 3.1 Water balance:

The general physical/chemical limnology of temporary aquatic habitats has been studied mainly in relation to shallow pools. Hartland-Rowe (1972) notes that the existence of any standing body of water, whether a lake, pond or pool, is the consequence of the balance between water gain and loss. For a permanent body, water input equals water loss. However, as we saw in Chapter 2, there are several sources of input (surface runoff, groundwater flow, precipitation, etc.) and several forms of output (absorption by the soil, evaporation, uptake by plant roots, etc.) the magnitude of which vary in space and time causing the water level to fluctuate. Water bodies in which input and output are highly variable but, none-the-less, often predictably cyclical, are invariably temporary. Many ephemeral streams, for example, experience several periods of flow each year and these may range in size from large flash floods to trickles. Even though their occurrence is erratic, over a period of years flow frequencies are predictable.

The exact length of the aquatic phase varies according to both geographic location and local hydrological conditions. For example, in the tropics, temporary pools and streams contain water immediately after the monsoons but soon lose water due to evaporation by the sun; in areas of extended high rainfall, the life of the water body will be extended until the dry season. In polar regions, although evaporation may not be as severe, a pond may dry up for a short period in late summer but may also "lose" its water in winter due to freezing solidly to its bed. The water balance in a temperate pond is illustrated in Figures 3.1 and 3.2. Sunfish Pond, Ontario, Canada, receives most of its water in spring from snow-melt. At this time, loss to surrounding soil is negligible as the pond margins are either saturated or still frozen. Evaporation is small also because of diminished solar radiation. As time passes, however, solar radiation intensifies, air and water temperatures rise and evaporation increases. Simultaneously, the groundwater table drops and the water level in the pond decreases. Eventually, in mid-summer, the pond becomes dry. Occasionally, very heavy rainfall may cause the pond to fill temporarily but

Figure 3.1: Sunfish Pond, Ontario, Canada in early spring (upper photograph) and in mid winter (lower photograph).

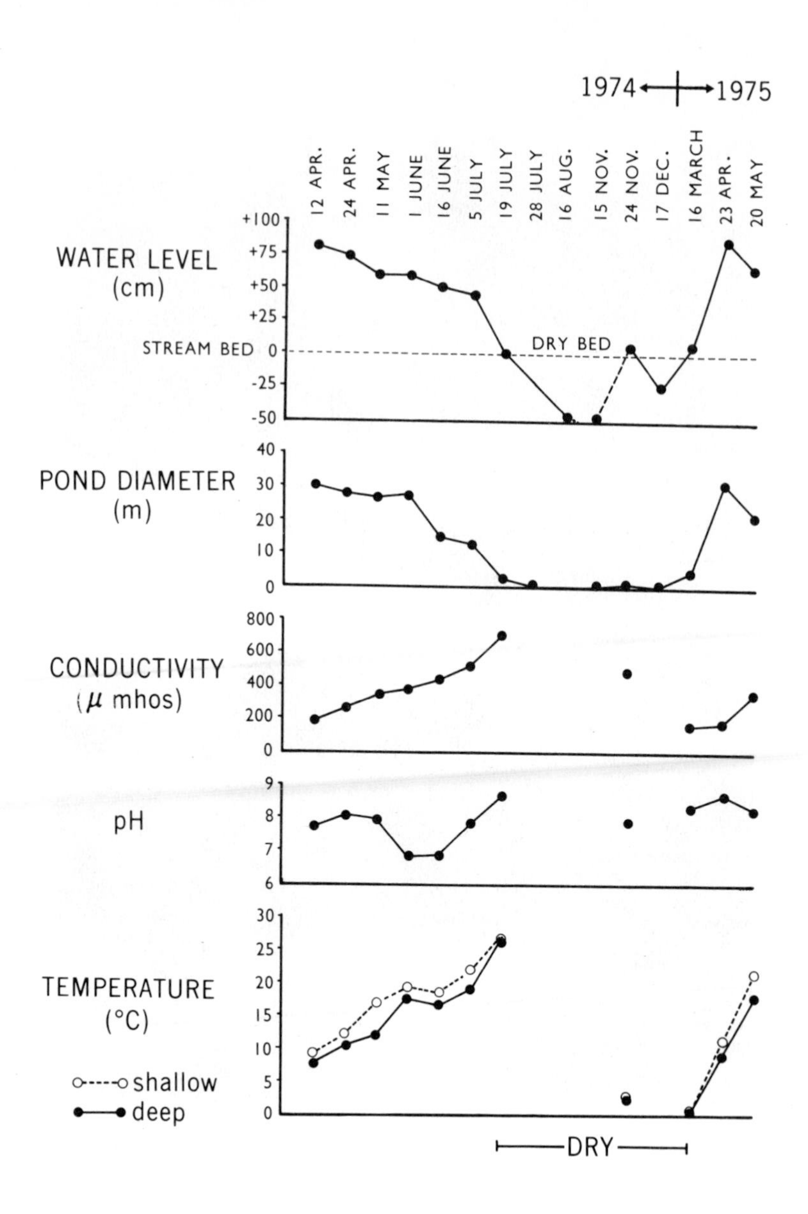


Figure 3.2: Environmental characteristics of Sunfish Pond.

usually the pond remains dry until the following spring. In years of low snowfall or early, warm spring, the life of the pond may be shortened. Conversely, in years of high snowfall or late, cold spring, it will be extended.

### 3.2 Water temperature:

Temperature is a very important variable, not just on a seasonal basis, but also on a daily or even hourly one. Because temporary waters are shallow, they are highly susceptible to heating up even to the bottom. The turbidity of the water is important in this respect, as highly turbid water heats up to a greater extent near the surface, whereas a clear pond will absorb much heat in its bottom mud leading to a more uniform water temperature. Pichler (1939) proposed a classification for small water bodies based on temperature, as follows:

1) *puddles* - small bodies up to 20cm deep with the bottom strongly heated by the sun; practically no stratification in the summer, when, daily, the variation may be as much as 25°C.

2) *pools* - water bodies up to 60cm deep, consequently less heat reaches the bottom; thermal stratification is upset daily by a turnover and the summer temperature variation may be up to 15°C at the surface and 5°C at the bottom.

3) *small ponds* - up to 100cm deep with very little heat reaching the substrate; stratification is more stable but can be upset daily, with summer temperature variation up to 10° at the surface and 2°C near the bottom. All these characters were based on open ponds and it must be realized that shading by emergent vegetation will make an important difference. In addition, a difference will be also evident between these pools and temporary streams, as the water in the streams is in motion - albeit slowly in some cases - and may run through shaded and non-shaded reaches.

The annual temperature regime of Sunfish Pond is shown in Figure 3.2. Temperature increases from a post-snow-melt value of around 8°C to a maximum of 27°C in mid-summer. Water in the deeper areas is typically 2-3°C cooler than surface water. There was little evidence of stratification in this clear-water pond.

Ponds that show stratification are frequently turbid and occur in regions where there is a marked difference between day and night air temperatures. As a result of daily turnover, some of these shallow water bodies experience, in 24 hours, thermal cycles that are annual in temperate lakes (Eriksen, 1966). This obviously affects the fauna.

To take an extreme example, surface water of shallow ponds in temperate regions may, on occasion, approach 40°C in mid-afternoon in summer. This is very near the thermal death point of most insects. Young and Zimmerman (1956) found many aquatic beetles active in such ponds in Florida even though the aquatic vegetation (*Chara* spp.) was dead. Frequently, however, these animals do not stay in the surface water, but concentrate under debris or burrow in the bottom mud, returning to the surface to forage at night and in the early morning.

In arctic and alpine regions, the effect of temperature is also marked but the conditions produced are quite different. Daborn and Clifford (1974) found that the first persistent ice cover in autumn, in a pond in western Canada, significantly reduced the normal close correlation between water and air temperatures. Initially, diurnal changes disappeared but subsequently reappeared as soon as the pond had frozen to the bottom. This was thought to be a function of lower thermal conduction across an ice-water interface than through ice alone and of variation in snow depth on top of the ice. Ice temperatures as low as -8°C were recorded and aquatic invertebrates, such as damselfly nymphs, survived embedded in the ice.

## 3.3 Turbidity:

As we have seen, turbidity may accentuate thermal stratification in a temporary pond. Frequently most of the absorption of solar heat occurs in the upper five or six centimetres below the surface (Figure 3.3). The source of turbidity varies. In the Oklahoma pond it was suspended clay particles as well as microscopic organisms. Chromogenic bacteria are responsible for the red colour of many briny desert ponds while purple sulphur bacteria often occur in shallow, stratified waters (Cole, 1968).

## 3.4 Dissolved oxygen and carbon dioxide:

Dissolved oxygen in temporary waters may fluctuate diurnally as a result of photosynthesis and respiration. Whitney (1942) found this oxygen pulse to be at a maximum just after dark when the day's photosynthesis had ended, but thereafter fell gradually due to overnight respiration. He concluded that, in many cases, absorption of oxygen from the air was of relatively minor importance, as often absorption values were far below the air saturation value for the particular temperature. Further, the oxygen content of the water frequently

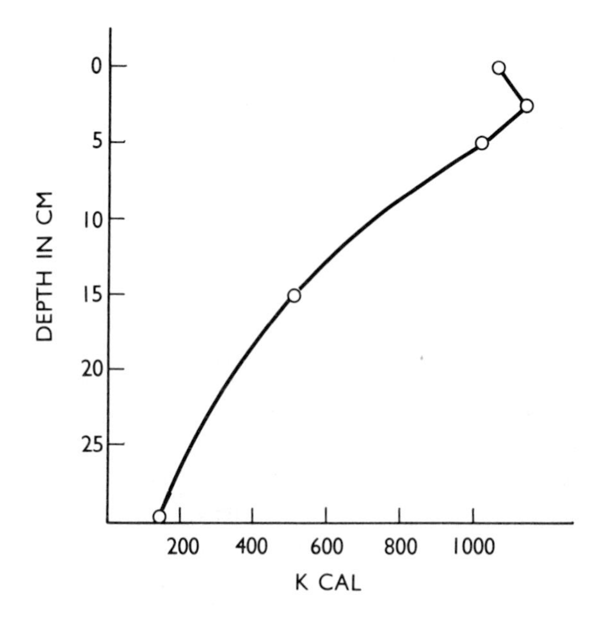


Figure 3.3: Heat absorption by one centimetre strata in a shallow temporary pond (180 ppm turbidity) in Oklahoma, U.S.A. (redrawn from Butler, 1963).

changed during a period when a uniform temperature prevailed. Schneller (1955) found that, during the low flow-stages of Salt Creek, Indiana, large quantities of decaying leaf matter were sufficient to cause an oxygen depletion combined with an increase in free carbon dioxide. Few mortalities amongst the fish species present were reported. George (1961), in a study of the diurnal variation of oxygen in two shallow ponds in India, found that, in one, the levels varied between 4.5 and 9.9ppm, being maximum at 5:30p.m. and minimum at 5:30a.m. In the other, the corresponding variation was from 0.1 to 28.2ppm. These increases were attributed to photosynthesis and the quantity of oxygen removed at night by community respiration (mostly *Mycrocystis aeruginosa*) was sufficient to cause a fish kill in one of the ponds. Cerny (cited in Vaas and Sachlan, 1955) found that in some small ponds in Europe, the amount of daily photosynthesis could completely exhaust all the available carbon dioxide. With this depletion of carbon dioxide, the pH could be expected to rise. But the magnitude of the change must depend not only on the intensity of the photosynthesis, but also on the degree of buffering available. Eriksen (op. cit.) showed that turbidity could cause stratification in the above parameters, as suspended material limits the penetration depth of sunlight thus restricting photosynthesis to the upper layers. This in turn might lead to oxygen, carbon dioxide and pH stratification.

### 3.5 Other chemical parameters:

The concentration and relative abundance of dissolved substances in temporary waters vary more than in most permanent waters. This is due, largely, to the physical processes to which temporary waters are subjected. I shall consider the changes associated with three common processes.

Figure 3.2 shows that in Sunfish Pond, Ontario, the conductivity of the water increased four-fold from the time that the pond filled in early spring to just before it dried up. This concentration of the chemical ions dissolved in the water was due to evaporation. Cole (1968) described the typical changes in water chemistry that occur as this proceeds - the normal trend is from carbonate to sulphatochloride to chloride water as calcium carbonate and, subsequently, calcium sulphate and possibly sodium sulphate are precipitated. The loss of calcium ions is accompanied by relative increases in sodium and magnesium ions. Some calcium remains, however, as calcium chloride - this salt being extremely soluble. Even this, though, may precipitate out of solution in ponds in extreme climates. Cole (1968) cites the example of Don Juan Pond in Antarctica. This is a calcium chloride pond, only 10cm deep, situated in a very dry, cold region. Needle-like crystals of calcium chloride hexahydrate form on the pond bottom and in the water.

In an intermittent stream in the desert region of Arizona, Grimm *et al.* (1981) showed that below-ground flow was an important source of nutrients, particularly nitrogen. They proposed a model to show that biological production in the stream should be greatest at points on the bed where subsurface water rises to the surface.

Concentration of chemicals in pond water may occur also as the result of freezing. Daborn and Clifford (1974), in a study of a shallow aestival pond in western Canada, found that as the winter ice cover thickened and the volume of the pond decreased, there was a rapid rise in conductivity, alkalinity, water hardness (calcium and magnesium ions), sulphate and orthophosphate. They found that this cryogenic "salting out" was a consequence of the stable and selective nature of ice crystals - only a few chemical elements or compounds have an appropriate configuration that allows them to become incorporated into the crystal structure of ice. They observed that the quantity of dissolved and particulate matter in the top layer of ice influenced the pattern of ice break up in the spring - primarily through control of the penetration of sunlight.

When temporary waters first fill after the dry phase, there may be a change in water chemistry, although this has not been as well studied. McLachlan *et al.* (1972) recorded the events as Lake Chilwa, a shallow, saline lake in Central Africa, filled after having been virtually dry for over a year. Refilling took five months during which time the authors observed a rapid dilution of initially high levels of dissolved salts, and a high suspended sediment load resulting from erosion of the lake bottom; this turbidity gradually decreased over a further two-year period. Increased turbidity is often evident also during the spring thaw of temporary ponds and streams in temperate regions.

Further examples of physical/chemical characteristics of specific temporary waters are given in the Case History section of Chapter 4 - The Biota.

# 4 THE BIOTA

## 4.1 The temporary water community - global scale:

### *4.1.1 Comparison of the communities of four ponds*

In this section we shall attempt to determine if there is such a thing as a community of organisms which can be readily identified as being universally characteristic of a temporary water body. To this end, in Table 4.1, I have listed the taxonomic groups recorded from temporary ponds in four well-separated regions of the world, eastern Canada, western Canada, northern Europe and New Zealand/Australia; species that overlap in two or more of the locations are named. This synthesis is provisional, as errors and omissions will result from incomplete identifications, insufficient collecting, endemism and local differences between pond types (e.g. particular water chemistry, temperature regime, riparian vegetation, etc.). Nevertheless, despite these shortcomings, considerable similarities in the faunas are evident and 14 species occur in two or more locations. Some, such as the snail *Physa fonticola*, the copepods *Eucyclops serratulus* and *Acanthocyclops vernalis*, and the beetle *Rhantus pulverosus* occur in both northern and southern hemispheres while others, such as the leech *Helobdella stagnalis*, the microcrustaceans *Canthocamptus staphylinoides*, *Daphnia pulex*, *Cypria ophthalmica*, *Cypridopsis vidua*, *Cypricercus ovum*, and the beetle *Anacaena limbata*, are holarctic in distribution. Other species are limited to one continent or locality.

Clearly certain taxa dominate the temporary pond community. Notable are the microcrustaceans, the mites and three insect groups, the true bugs, the beetles and the midges or gnats. Species within these groups invariably show special characteristics of either their physiology or life cycle which make them successful in temporary waters as well as perhaps allowing them the means to colonize them. We shall defer discussion of this topic to a later section of this book. Similarly, there are certain groups of aquatic invertebrates that do not occur in temporary ponds. These include the sponges, the alderflies, the stoneflies and the caseless caddisflies. In the case of the latter two groups, however, this may be the result of the habitat being a lentic (standing water) one rather than an intermittent one, as both have been recorded from temporary streams. The indication from this

Table 4.1: Faunal overlap between temporary ponds in four regions of the globe. (*indicates presence of similar genera or ecological equivalents; identical species are named)

| Taxa | Ontario | Vancouver Island | Hamburg | New Zealand/ Australia |
|---|---|---|---|---|
| TURBELLARIA (flatworms) | * | * | * | * |
| NEMATODA | * | * | * | * |
| ROTIFERA | – | – | * | * |
| OLIGOCHAETA (true worms) | *Lumbricus variegatus* | *L. variegatus* | *L. variegatus* | * |
| HIRUDINEA (leeches) | *Helobdella stagnalis* | – | *H. stagnalis* | – |
| GASTROPODA (snails) | * | * | *Physa fontinalis* | *P. fontinalis* |
| BIVALVIA (clams) | * | – | * | – |
| TARDIGRADA (water bears) | – | – | * | – |
| ECTOPROCTA (moss animals) | * | – | * | * |
| ANOSTRACA (fairy shrimp) | * | * | – | * |
| CONCHOSTRACA (clam shrimp) | * | – | – | * |
| NOTOSTRACA (tadpole shrimp) | – | – | – | * |
| CLADOCERA (water fleas) | *Daphnia pulex* | *D. pulex* | *D. pulex* | * |
| COPEPODA | *Acanthocyclops vernalis* *Canthocampus staphylinoides* | *A. vernalis* *C. staphylinoides* | *Eucyclops serratulus* | *A. vernalis* *E. serratulus* |
| OSTRACODA | *Cypria ophthalmica* *Cypridopsis vidua* | *C. ophthalmica* *Cypricercus ovum* | *C. ophthalmica* *C. vidua* *C. ovum* | * |
| AMPHIPODA (side-swimmers) | * | – | – | – |
| ISOPODA (sow-bugs) | * | * | * | – |
| DECAPODA (crayfish) | * | – | – | – |
| ACARI (mites) | *Thyas barbigera* | *T. barbigera* | * | * |
| COLLEMBOLA (springtails) | * | * | * | – |

Continued

Table 4.1: Faunal overlap between temporary ponds in four regions of the globe. (*indicates presence of similar genera or ecological equivalents; identical species are named)

| Taxa | Ontario | Vancouver Island | Hamburg | New Zealand/ Australia |
|---|---|---|---|---|
| EPHEMEROPTERA (mayflies) | * | – | – | – |
| ODONATA (dragonflies) | * | * | * | * |
| HEMIPTERA (true bugs) | * | * | * | * |
| COLEOPTERA (beetles) | *Anacaena limbata* | * | *A.limbata Rhantus pulverosus* | *R.pulverosus* |
| TRICHOPTERA (caddisflies) | *Limnephilus indivisus* | *L.indivisus* | * | * |
| DIPTERA (true flies) | * | * | * | * |
| CHIRONOMIDAE (midges or gnats) | * | * | * | * |
| AMPHIBIA | * | * | * | * |

Sources: Stout, 1964; Barclay, 1966; Bishop, 1974; Williams, 1975; Morton & Bayly, 1977; Wiggins *et al.,* 1980; Caspers & Heckman, 1981; Bayly, 1982; Williams, D.D., 1983.

table is that temporary ponds, throughout the world, provide very similar niches for colonizing animals. In many instances, these niches are filled by the same genera, lending credence to the theory that the taxonomic unit of "genus" is an ecological as well as a morphological entity (Wiggins and Mackay, 1978). In the case of cosmopolitan, readily-disseminated forms, such as *Daphnia pulex*, these niches are filled by identical species or members of a species complex. Where dispersal powers are weak and do not allow a species to colonize habitats far afield, locally endemic species of the same major taxon fill the gap. Thus, for example, in Ontario temporary ponds the single fairy shrimp species is *Chirocephalopsis bundyi*, while in the Vancouver Island pond it is *Eubranchipus oregonus* and in Australia it is *Branchinella australiensis*.

### *4.1.2 Distribution patterns in the Branchiopoda*

Although there is a general dearth of comparative data on temporary water communities, some individual components of the communities are better known. One such group is the Branchiopoda, and the Anostraca (brine shrimps and fairy shrimps) in particular. As we have seen already, the Anostraca are important members of the communities of temporary lentic waters throughout the world and, later, we shall be examining the adaptations that allow them to live in these habitats.

A more detailed pattern of their global distribution than was possible in Table 4.1 shows that the Anostraca occur on all continents (Table 4.2). The majority (17) of present-day genera (23) appear either to have evolved since the formation of the modern continents or represent relict groups. Resolution of these alternatives is hampered by the paucity of their fossil record (Tasch, 1969). Modern distribution patterns that suggest extensive distribution in Pangea (the single super-continent that existed in the Paleozoic Era, some 200 million years ago) are seen in only three genera - *Artemia, Branchinella* and *Branchinecta*. Three other genera, *Artemiopsis, Linderiella* and *Streptocephalus,* have current distributions that suggest that they were widely distributed across Laurasia (the northern landmass that was derived from the breakup of Pangea in the late Triassic Era, some 180 million years ago; Figure 4.1). Belk (1981) proposed that many of the presumptive Pangea genera do not occur, now, on all the modern continental fragments of Pangea because of ecological factors. *Branchinecta*, for example, is a genus of cold-water adapted species that is particularly common in the Canadian Arctic and Alaska. It is absent from Australia which, when part of the east coast of Pangea, had a warm climate (Bambach *et al.*, 1980). Similarly, *Artemiopsis*, a presumptive Laurasian cold-water genus, is restricted to cold northern regions of Eurasia and North America (e.g. the Northwest Territories of Canada). Conversely, *Branchinella* and *Artemia*, which are warm-water forms, are absent from modern-day Antarctica. The limited distribution of *Artemia* in Australia may be due, in part, to competition from Australia's native brine shrimp, *Parartemia* (Geddes, 1980). The absence of streptocephalids from Australia and South America suggests that *Streptocephalus* (another presumptive Laurasian genus) colonized Africa after the breakup of Gondwanaland and has since undergone a major adaptive radiation there - though mostly in the temperate zone (Brtek, 1974; Belk, 1981).

Table 4.2: World distribution of the 23 genera of Anostraca (after Belk, 1981)

| | Eurasia | Africa | North America | South America | Australia | Ant-arctica |
|---|---|---|---|---|---|---|
| *Branchinecta* | * | * | * | * | | * |
| *Artemia* | * | * | * | * | * | |
| *Branchinella* | * | * | * | * | * | |
| *Streptocephalus* | * | * | * | | | |
| *Linderiella* | * | * | * | | | |
| *Branchinectella* | * | * | * | | | |
| *Branchipodopsis* | * | * | | | | |
| *Branchipus* | * | * | | | | |
| *Tanymastix* | * | * | | | | |
| *Chirocephalus* | * | * | | | | |
| *Artemiopsis* | * | | * | | | |
| *Drepanosurus* | * | | | | | |
| *Polyartemia* | * | | | | | |
| *Siphonophanes* | * | | | | | |
| *Tanymastigites* | | * | | | | |
| *Metabranchipus* | | * | | | | |
| *Dexteria* | | | * | | | |
| *Eubranchipus* | | | * | | | |
| *Polyartemiella* | | | * | | | |
| *Thamnocephalus* | | | * | * | | |
| *Dendrocephalus* | | | | * | | |
| *Phallocryptus* | | | | * | | |
| *Parartemia* | | | | | * | |

Belk (1981) provided evidence for his idea that climate is an important controlling factor in anostracan distribution by comparing the anostracan faunas of Arizona and South India, two climatically different regions. Arizona is a region in the temperate zone that experiences marked seasonal differences in climate and includes environments which range from lowland desert to high alpine. The anostracan fauna is represented by 13 species from 5 genera (*Artemia, Branchinecta, Eubranchipus, Streptocephalus* and *Thamnocephalus*). South India is a region in the tropical zone that experi-

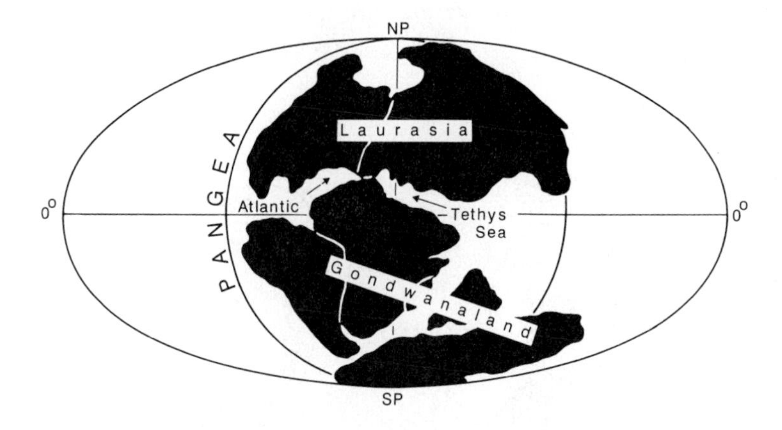


Figure 4.1: Map showing the breakup of Pangea in the late Triassic Era (180 million years B.P.).

ences only slight seasonal changes in climate. The area of the Western Ghats mountains and to the west is humid and anostracans are not found there, but east of the mountains is a semiarid area where temporary ponds are very common. Here the anostracan fauna is represented by only 6 species from 3 genera (*Artemia, Branchinella* and *Streptocephalus*). The variation in thermal regime associated with refilling of temporary ponds is much less in Southern India than in Arizona and Belk maintained that because there are significant differences between the temperature requirements for egg hatching between species, a lower diversity is to be expected in the less variable environment. Anostracan species richness, in general, is always highest in temperate parts of the globe.

There are global and regional differences in the distribution of the Notostraca (tadpole shrimps) also. Rzoska (1984) stated that the genus *Lepidurus* is northern and may occur in Arctic conditions, whereas the genus *Triops* (=*Apus*) is confined to warmer waters, especially in arid regions and hot climates. In the United States, the single species of *Triops* (*T. longicaudatus*) seems to be confined to the drier western half of the country, and its range extends down into Mexico, the West Indies, the Galapagos Islands, Argentina and the Hawaiian Islands. There is one record for Canada, in Alberta (Figure 4.3). The genus *Lepidurus* is represented by several species in North America and these again occur chiefly in the western states but also in central and arctic Canada. *Lepidurus arcticus*, (Figure 4.2), for example, is

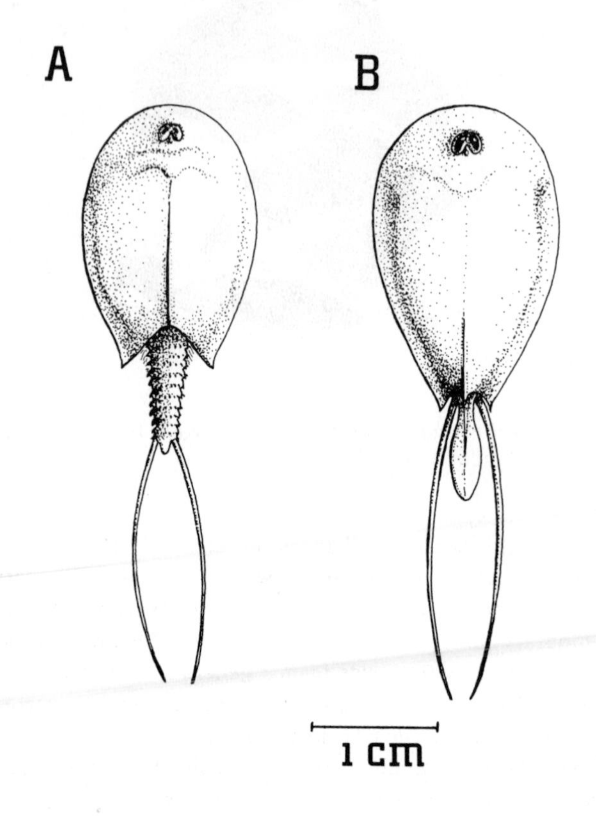


Figure 4.2: Notostraca: A – *Lepidurus arcticus* Pallas 1793
B – *Lepidurus couesii* Packard 1875

found only in Alaska, the Northwest Territories and Labrador, plus in Greenland (Figure 4.3). *Lepidurus couesii* (Figure 4.2) is found in Montana, North Dakota, Oregon, Idaho and Utah in the U.S.A.; Alberta, Saskatchewan and Manitoba in Canada; and in Russia, Northern Siberia and Turkestan in the Palearctic (Linder, 1959; National Museum of Canada records).

In Australia, both *Lepidurus* and *Triops* occur in all states ex-

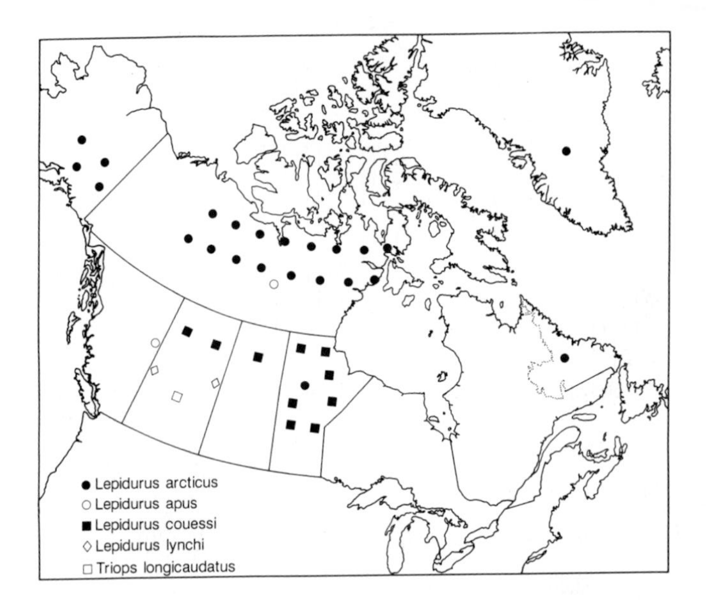


Figure 4.3: Records of notostracans from Canada, Alaska and Greenland (dots simply represent single collections made in a particular province or state – they do not accurately depict where in the province/state the collections were made. Based on records in the National Museum of Natural Sciences, Ottawa, and in Linder, 1959).

cept Tasmania, where only *Lepidurus* is known; only *Triops* has been found in the Northern Territories. However, *Lepidurus* is largely confined to the more temperate southeastern and southwestern corners of the continent, while *Triops* tends to be more common in the dry interior, although it also occurs on the coast in a few regions (Figure 4.4). These distribution patterns seem to be correlated with regional differences in climate, particularly in terms of mean annual temperature and evaporation (Figure 4.5). Even though the ranges of these two genera overlap in the southeast, they are never found coexisting in the same body of water and this also appears to be true for much of the rest of the world (Williams, 1968).

The Conchostraca (clam shrimps) have a wide geographic distri-

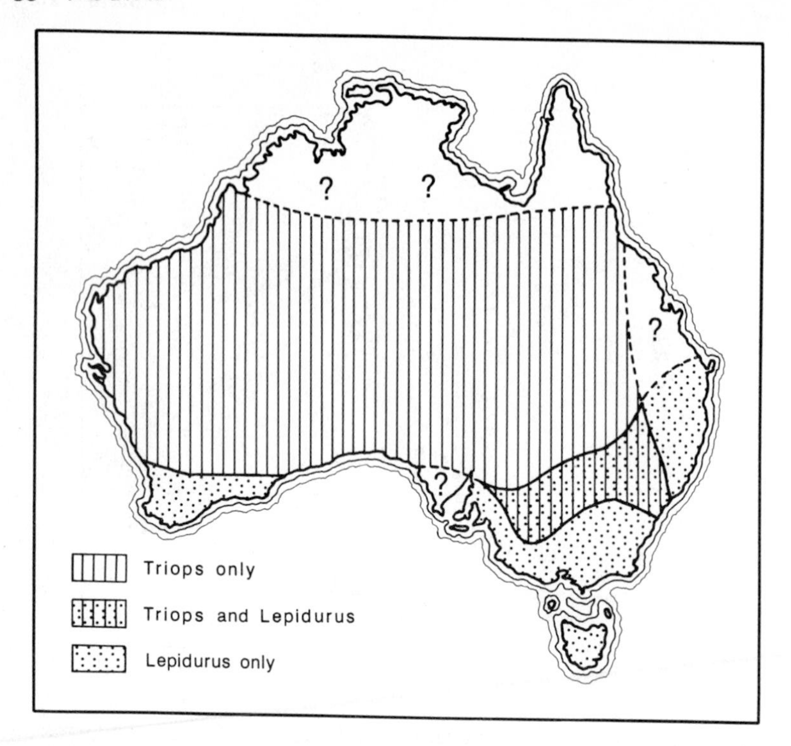


Figure 4.4: Geographical distribution of *Triops* and *Lepidurus* (Notostraca) in Australia (redrawn from Williams, 1968).

bution primarily in temporary waters but also in the littoral zone of lakes. Some species have very extensive distributions while others are known only from their type locality. For example, in North America, *Lynceus brachyurus* is found across the United States as well as in most Canadian provinces (excepting the Maritimes), the Northwest Territories and Alaska (Figure 4.6). It is found also in Europe and Asia. *Cyzicus mexicanus* occurs across most of the U.S. and Mexico, and also in Alberta and Manitoba (Mattox, 1959; National Museum of Canada records). Some species of *Eulimnadia*, however, have very restricted distributions. *Eulimnadia alineata* and *E. oryzae*, for example, have been recorded only from rice fields at Stuttgart, Arkansas (Mattox, 1959). There are, nevertheless, three species of *Eulimnadia* recorded from Australia (from the Northern Territories, central and Western Australia, Queensland and Tasmania). There are some 23 species of conchostraca in Australia which are all, with the exception

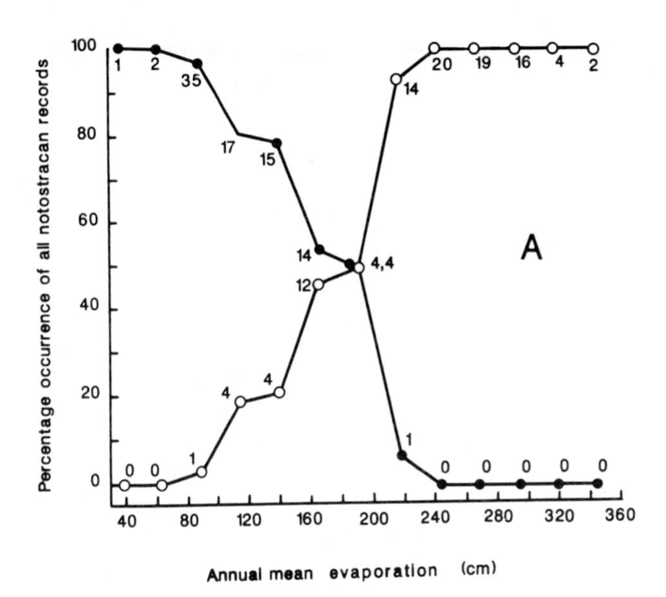


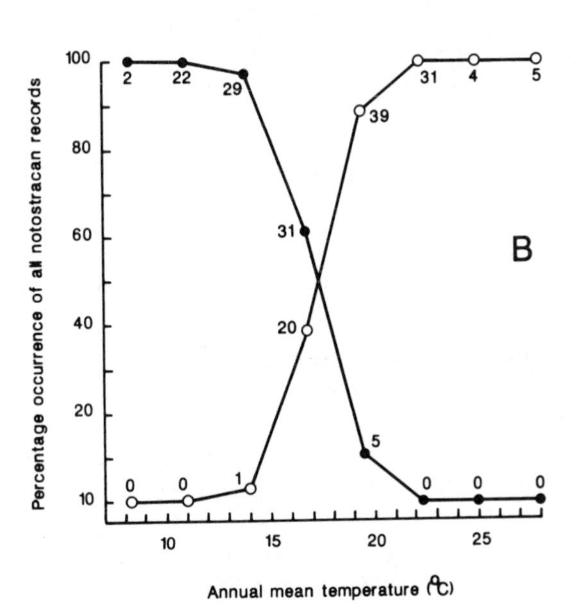


Figure 4.5: Distribution of *Triops* (open circles) and *Lepidurus* (closed circles) (Notostraca) with respect to: A – isoclines of annual mean evaporation; B – isotherms of annual mean temperature (numbers of records are given; redrawn after Williams, 1968).

of *Cyclestheria hislopi*, endemic (Williams, W.D., 1980). In general, the Conchostraca are typically found in warmer waters than most of the Anostraca.

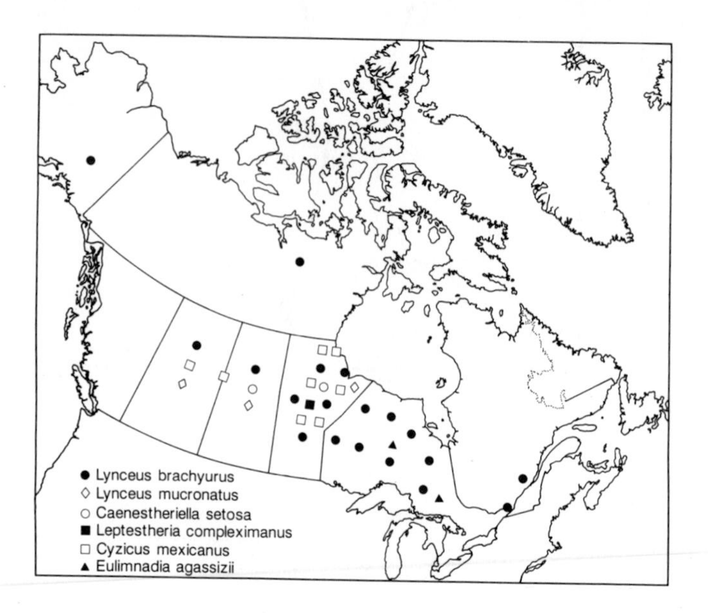


Figure 4.6: Records of conchostracans from Canada and Alaska (dots simply represent single collections made in a particular province — they do not accurately depict where in the province the collections were made. Based on records in the National Museum of Natural Sciences, Ottawa, and in Mattox, 1959).

## 4.2 The temporary water community - local scale - case histories:

Temporary waters vary considerably throughout the world because of differences in local climate, topography, geology, etc. Since there are differences in their biotas because of these environmental factors and distributional characteristics of species (e.g. endemism versus cosmopolitanism), it is not possible to provide a detailed comparison of all habitat types (as was attempted for ponds in temperate regions of the world in Section 4.1.1) in a uniform style. In this section, therefore, we shall examine, on an individual case history basis, a cross-section of those temporary waters that have been studied in detail.

### 4.2.1 *Biota of a nearctic temporary pond*

Sunfish Pond, Ontario, Canada, is a typical temperate, temporary vernal pool. This categorization was established by Wiggins (1973) and indicates a pool which derives its water primarily from rain and melting snow in early spring and which becomes dry in early summer, leaving a water-free basin for 8-9 consecutive months of the year, including winter. Ponds that retain water in the autumn, winter and spring, and only become dry for 3-4 months in the summer, are termed temporary autumnal pools.

*Community succession*

98 taxa have been identified from Sunfish Pond. Figure 4.7 shows some examples of these species arranged, not by taxonomic groups, but according to their seasonal occurrence in the pond. A succession is evident which, at first glance, seems fairly continuous but which can be divided into several distinct faunal groups.

Group 1 contains animals that can be found during virtually the entire aquatic phase of the pond. During the dry phase they can be dug up from the substrate as semi-torpid adults or immature stages. If placed in water they become mobile within minutes. Included in this group are the clams and snails, the aquatic worms, a beetle, a copepod and a haemoglobin-containing midge.

Group 2 contains taxa present as active forms within a few days of the pond filling in the spring. These species mostly complete their life cycles within 4-6 weeks and disappear (by entering a resting stage - usually as eggs or diapausing immatures - or by leaving the pond as emerged adults) well before (4-6 weeks) the pond dries up. This, the largest group, contains the microcrustaceans, fairy shrimp, mites, mosquitoes, cased caddisflies, bugs, some midges, beetles and a dragonfly.

Group 3 contains taxa which appear 2-5 weeks after pond formation in the spring. It includes a conchostracan (clam shrimp), a damselfly, midges and beetles, some of which appear only as adults and do not breed in this pond. Life cycles of species in Group 3 are typically completed in 5 weeks.

Group 4 taxa appear 2-3 weeks before the pond dries up (approximately 10 weeks after filling). The taxa include beetles, mayflies

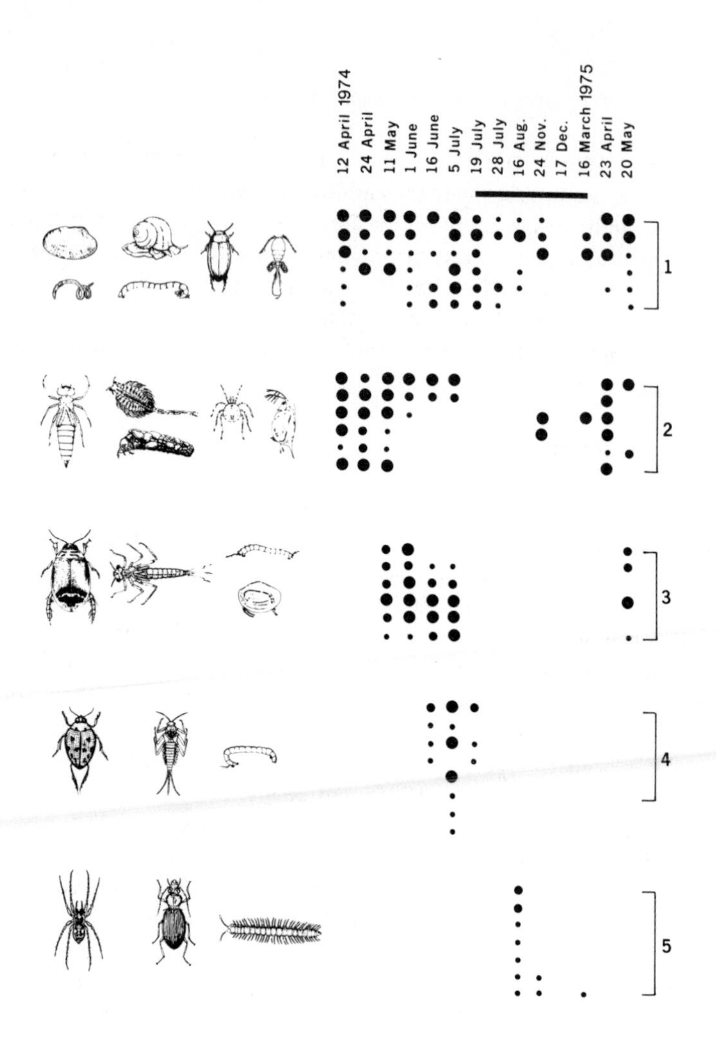


Figure 4.7: Seasonal succession of the fauna of Sunfish Pond (duration of the dry phase is indicated by the horizontal black bar; circles indicate relative abundance on a decreasing scale of "abundant, common and rare"; faunal groupings referred to in the text are indicated on the right-hand side of the figure).

and midges.

Group 5 contains taxa which appear only in the dry phase. These are primarily terrestrial or riparian species and include millipedes, centipedes, spiders and beetles.

Figure 4.8 shows the seasonal variation in the total number of

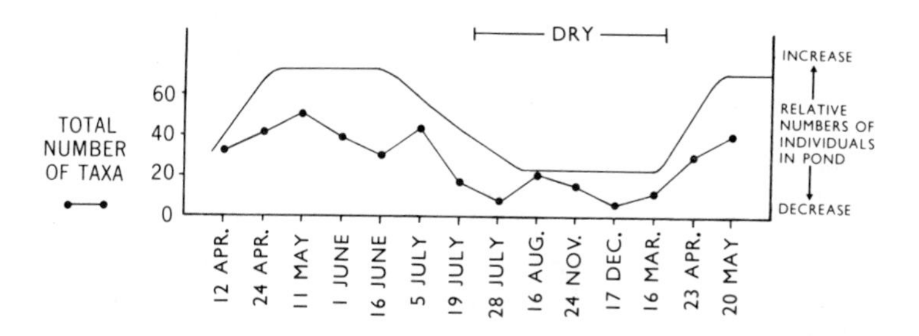


Figure 4.8: Seasonal variation in the total number of taxa (solid circles) seen in Sunfish Pond, together with an estimate of the relative numbers of individuals present (solid line).

taxa seen in the pond. Diversity is greatest during the wet phase of the habitat, but the drying basin by no means lacks a fauna.

### *4.2.2 Biota of a nearctic temporary stream*

Whereas the differences between permanent standing and running water may be very considerable, in the temporary water situation, because many of these streams flow slowly and often dry up to form pools, we find considerable overlap in the faunas. However, it is important to realize that the lotic (running water) phase of the habitat adds to the diversity of niches available.

Moser Creek, Ontario, the example we shall consider, is a small 400m long by 80cm wide, temporary stream which flows for about 6 months of the year (November-April). During May and early June it consists of a series of unconnected pools which dry up completely by July.

*Community succession*

The animals found in this stream can be divided into three successive groups based on the water conditions found in the habitat throughout the year. The major components of these three groups are summarized in Figure 4.9. Inclusion in a particular group indicates that this is where the species passes the most active stages of its life - undoubtedly there will be some overlap.

| Fall-Winter Stream Fauna | Spring-Pool Fauna | Summer-Terrestrial Fauna |
|---|---|---|
| Hydras | Water fleas | Terrestrial worms |
| Flatworms | Seed shrimps (some) | Slugs |
| Aquatic worms | Mayflies | Terrestrial snails |
| Seed shrimps (some) | Dragonflies | Mud-loving beetles |
| Copepods | Aquatic Bugs | Rove beetles |
| Amphipods | Craneflies (some) | Ground beetles |
| Crayfish | Mosquitoes | Ants |
| Stoneflies | Shore flies | Scavenger flies |
| Caddisflies | Rat-tail maggots | Dung flies |
| Crane flies (some) | Moth flies | Biting midges |
| Blackflies | | Spiders |
| Midges | | |

Figure 4.9: Faunal succession in a Nearctic temporary stream.

The fall-winter stream fauna consists of those animals which appear shortly after the stream starts flowing in the autumn and most reproduce successfully before the flow stops in the spring. Many of them, for example some of the midges and the cased caddisflies, grow quickly before the water cools down and then are ready, shortly after ice breakup, to pupate, emerge, mate and oviposit. Other midges and the amphipods grow steadily throughout the winter and mature a little later. A few midge species grow so slowly that their period of activity spans two groups.

The spring-pool fauna represents those species that wait until the stream has stopped flowing and only shallow pools remain. These pools are excellent breeding environments due to the ease with which they warm up and the abundant plant food that develops in them. There are two basic categories of species that use these pools. The first, which includes mosquitoes and some of the aquatic beetles, has been present since the previous autumn (the mosquitoes as eggs and the beetles as adults), but they do not produce any larvae until the flow stops. Once hatched, the larvae grow quickly and mature before the pools dry up. The second category consists of other beetles and the water striders. These fly or walk in as adults just as the pools are forming, lay their eggs and generally leave. Their larvae quickly hatch and mature just as the pools dry up. Occasionally, an unusually dry spring may speed up evaporation so that the life cycle is not complete when the water disappears. Some of the species in this group, however, have pupae that are capable of resisting drought for a short while, thus enabling the adults to emerge successfully.

The summer-terrestrial fauna consists mostly of riparian species that move onto the streambed once it has dried. Some, such as the earthworms, are probably attracted by the dampness while others, like the slugs and snails, come to feed on the exposed algal mats. The beetles and ants scavenge the bed for dead and dying individuals left over from the aquatic phase and the spiders, in turn, follow these, their prey. A few specialized taxa belong in this group also, for example, dipteran flies of the families Sepsidae and Sphaeroceridae; these flies are generally attracted to damp, decaying material on which they lay their eggs.

Note that certain major taxa are very definitely characteristic of the three phases and the similarity of the fauna of the spring-pool phase to that of a temporary pond. There tends, however, to be a substantial reduction in the diversity of the microcrustaceans (Ostracoda, Copepoda and especially Cladocera) and euphyllopod crustaceans (fairy shrimps, clam shrimps and tadpole shrimps) in temporary streams. Their absence may be due to dominance of the running water phase of the habitat or, in the case of the latter, to the presence of fishes which often migrate into temporary streams. Other, macrocrustaceans, particularly amphipods and isopods, are common in temporary streams. Notable additions to the temporary stream fauna are certain species of stoneflies, blackflies and midges, most of which require some sort of current. Greater details of the life cycles of some of these species will be given in Chapter 5.

### 4.2.3 *Australian temporary streams*

There are few data available on the abiotic features of temporary streams in Australia despite the widespread occurrence of these habitats across this continent; over half of the land mass is drained by intermittent streams and rivers (Williams, W.D., 1983). The data that do exist indicate the same, wide fluctuations in factors like pH, dissolved oxygen, temperature and conductivity as have been noted in the Northern Hemisphere (Boulton and Suter, 1986).

Similarly, only a few studies of the fauna exist but these suggest that Australian temporary streams are perhaps richer in species than their counterparts elsewhere. Further, species richness appears to increase with increasing permanence of the water. Most of the macroinvertebrate community is comprised of insects (>75%) with the Diptera (chiefly tipulids, chironomids, simuliids and ceratopogonids) being dominant. Other insect groups present are the Coleoptera (dytiscids and hydrophilids), Trichoptera (leptocerids and hydrobiosids in Victoria and South Australia), Plecoptera, Hemiptera, Odonata and Ephemeroptera, in roughly decreasing order of proportional representation. Regional differences are apparent (e.g. the plecopteran fauna of South Australia is relatively poor) and this may explain the fact that the proportions of various insect groups represented are somewhat different from those in Northern Hemisphere streams of similar latitude (Boulton and Suter, 1986).

During their terrestrial phase, two temporary streams in Victoria were seen to be colonized by a "clean-up crew" not unlike that reported for Moser Creek, Canada (Section 4.2.2). The Victorian fauna was comprised of carab and hydraenid beetles, lycosid spiders, ants and terrestrial amphipods (Boulton and Suter, 1986).

Temporary streams running through *Eucalyptus* forest receive a distinct peak in input of leaf litter during summer and this coincides with the period of low or zero flow (Lake, 1982). This is somewhat different from temperate Northern Hemisphere streams where most of the input of deciduous leaves is in the late autumn, a time when some temporary streams are flowing. When flow began in Brownhill Creek in South Australia, Towns (1985) recorded a pulse of organic matter, consisting of both coarse particles and dissolved materials, which was carried downstream. In Australia, as elsewhere, little is known, quantitatively, of the fate of detritus and its derivatives, their partitioning in the foodweb, and their importance to the energy budget of the system compared with, say, autochthonous input (Boulton and Suter, 1986).

### 4.2.4 *Billabong systems*

Australia is, climatically, an arid continent exhibiting both low levels of precipitation and high evaporation rates. Australian drainage basins reflect these features in that many of them show wide variation in streamflow on both seasonal and annual time scales (Williams, 1981a). Intermittent flow and high evaporation rates cause water levels in entire river or pond systems to fluctuate sufficiently to produce a biota adapted to cyclical loss of water. Such temporary waters are to be found in all parts of the continent but the example chosen here is Magela Creek, a seasonally-flooding tributary of the East Alligator River in the Northern Territory, about 250km east of Darwin.

Magela Creek is in a region subject to a tropical monsoon climate and thus has distinct wet and dry seasons with 97% of the 153cm annual rainfall occurring from October to May. During the wet season, the Magela floodplain is a continuous body of water covering some $190km^2$ to depths of between 2 and 5m and water velocities may exceed 1m/s. During the dry season, it consists of a series of discrete billabongs, or lagoons, of varying size (Morley *et al.*, 1985).

These billabongs undergo seasonal and diurnal fluctuations in both physical and chemical properties of their water. Surface water temperatures vary between 22$^{o}$C in July and 39$^{o}$C in November and diurnal fluctuations are greatest in the dry season. Turbidity tends to increase throughout the dry season but is less in the wet season. Conductivity increases during the dry season and pH varies between 6.0 and 7.0. Oxygen levels vary according to temperature and the amount of photosynthetic activity of macrophytes and phytoplankton but at no time is there complete deoxygenation (Marchant, 1982a). The billabongs in the Magela System can be subdivided into two basic types: (1) those on the floodplain which are separated from the creek by a levee but which have an intermittent connection with the creek thus enabling them to fill and drain - these have been termed backflow billabongs; (2) channel billabongs which occur on the main channel of the creek and have separate inlets and outlets (Walker and Tyler, 1979). The former are usually shallower than the latter and there are consequently some differences between the types of environments that they present for the biota. Marchant (1982a) recorded some differences in the composition of the littoral faunas between these two types of billabong but chiefly among the less common taxa. The littoral fauna appears to be rich in species with particularly high densities of Ephemeroptera, Trichoptera, Mollusca, Hemiptera and Chironomidae occurring in macrophyte beds during the wet season. Taxa predominant in the dry season include Coleoptera (especially adult Dytiscidae), tanypodine chironomids, Ceratopogonidae, some Hemiptera, some Gastropoda (e.g. *Ferrissia* and the prawn *Macrobrachium*. Less common taxa found at different times of the year include Tricladida, Oligochaeta, Hirudinea, Hydracarina, Porifera, Hydridae, Gordiidae, Ostracoda and Conchostraca.

In the shallow Magela billabongs the greatest densities and diversities of animals occur during the late wet season - early dry season with as much as a five-fold factor difference in the numbers of individuals compared with other times of the year. Seasonal fluctuations in density and diversity do not appear as marked in the channel billabongs. Temporal fluctuation is probably the result of changes in macrophyte abundance, and maximum animal biomass coincides with maximum plant biomass (April-July). The plants provide both food for macroinvertebrates and protection from predators. It is likely, as in many other aquatic habitats, that the macrophytes are not eaten directly but after they have died and become part of the general pool of detritus on the pond bottom. Epiphytic algae on the live macrophytes may be another important source of food.

The fauna survives the dry season by a variety of methods including hibernation in the bottom mud - Gastropoda; resistant eggs - Ephemeroptera; and recolonization from other billabongs, particularly those in the main channel which are deeper and therefore virtually all permanent. In fact, during the few years that the Magela System has been studied, none of the main billabongs have dried up totally. The beginning of the wet season is characterized by a rapid resurgence of the fauna, especially in the shallow billabongs, and many species appear to have short life cycles (e.g. circa one month) with fast rates of larval growth (Marchant, 1982b).

The general features of the fauna of the Magela billabongs parallel those known for similar habitats (e.g. floodplain river systems; see Welcomme, 1979, and Chapter 8) in other regions of the tropics, and these habitats support at least as many species as much larger temperate Australian lakes. Outridge (1986) has argued that the high species diversity found in the Magela communities is due to rarefaction and predictable environmental heterogeneity, related to the monsoonally-influenced variations in flow and water quality.

### *4.2.5 Sahara Desert rainpools*

In the whole of northern Sudan and southern Egypt, there are no standing waters other than those created by rain which collects in depressions in the Nubian Desert (Rzoska, 1984). These rainpools are several hundred square metres in area but are no more than 50cm deep. The area is subject to high air temperatures and severe winds, and the rainfall itself is irregular. Around Khartoum, there are typically eight dry months in a year with perhaps four to ten episodes of rain during July, August, September and October. Mean annual rainfall is between 160 and 180mm, and only episodes of greater than 15mm form rainpools. As soon as they are formed, the hot dry air (28-41$^{o}$C) and winds cause rapid evaporation such that the typical duration of these pools is only 7-15 days.

The main elements in the fauna of these pools, recorded by Rzoska, are crustaceans, particularly the Branchiopoda (Table 4.3). In

Table 4.3: The crustacean fauna of temporary rainpools near Khartoum (after Rzoska, 1984).

| | |
|---|---|
| Anostraca: | *Streptocephalus proboscideus* |
| | *Streptocephalus vitreus* |
| | *Branchipus stagnalis* |
| Conchostraca: | *Eocyzicus klunzingeri* |
| | *Eocyzicus irritans* |
| | *Leptestheria aegyptiaca* |
| | *Limnadia* sp |
| Notostraca: | *Triops granarius* |
| | *Triops cancriformis* |
| Cladocera: | *Moina dubia* |
| Copepoda: | *Metacyclops minutus* |
| | *Metadiaptomus mauretanicus* |

addition there are nematode worms, protozoans, chironomid larvae and other adult insects, and two rotifers, *Asplanchna* and *Pedalion*.

With such a short-lived habitat, the fauna is highly specialized and shows rapid development. The Conchostraca (clam shrimps), for example, appear in the Khartoum pools within three days of pool formation, and are mature and bearing eggs by day five. The waterflea, *Moina dubia*, is fully grown within 72 hours and breeds on the third and fourth days, first parthenogenetically and also by ephyppia (see Chapter 5). The copepod, *Metacyclops minutus* is fully grown within 48 hours and then breeds immediately. High population densities are achieved rapidly, reaching maxima of 460 *Moina dubia*/litre of water, and 800 *Metacyclops minutus*/litre, on top of which there are many more immature individuals (up to 1700/litre). The larger species, such as the notostracans, *Triops* spp. (tadpole shrimps), appear in the pools on day seven.

Rzoska (1984) could only speculate on the nature of the food chains in these pools. Based on other studies it is known that the Anostraca (fairy shrimps) and Conchostraca are filter feeders, that *Triops* is a scavenger and a carnivore, and that the microcrustaceans are fine-particle feeders. Rzoska suggested that protozoans and bacteria might form the basis of the trophic pyramid, although he did, once, observe a bloom of the blue-green alga *Oscillatoria*.

### *4.2.6 Rain-filled rock pools of African "Kopjes"*

Shallow depressions in bedrock which periodically collect rainwater are common features in tropical and subtropical regions. On isolated hillsides in tropical Africa, McLachlan and Cantrell (1980) identified three types of such pools. They differ mainly in depth which causes a gradation in life span following rain; none, however, contain water during the six month dry season. The relative life spans of these pool types are: *A* - approximately 24 hours; *B* - several days; *C* - several weeks. The faunas of these pools are dominated by dipteran larvae which can achieve high densities (50-300 $\times 10^3$ larvae/pool) but which appear highly specific to a particular pool type. Pools of type A are populated by larvae of the chironomid *Polypedilum vanderplanki* which is physiologically very tolerant of drought conditions; type B pools are inhabited by larvae of the biting midge *Dasyhela thompsoni* which bury themselves in the mud of the pond bottom when the water evaporates; while type C pools, the longest-lived, are populated by larvae of the chironomid *Chironomus imicola* which have little tolerance of desiccation.

Food in these pools is generally abundant and is typically allochthonous in origin (i.e. it comes from outside the pools); it consists of fruit, pollen, flowers and the dung of small carnivores (McLachlan, 1981). There are few predators in the aquatic phase of these habitats but dormant stages are susceptible to scavenging by terrestrial taxa, such as pheidolid ants, particularly when the water has just evaporated but the mud is still damp. Once the mud dries totally, predation ceases.

Many of these rockpools support virtual monospecific populations. This might be explained in terms of the rigours of the environment or perhaps some other factor; the evidence will be discussed in the chapter on Colonization Patterns.

### *4.2.7 Rain-filled rock pools of southwestern Australia*

A series of shallow, water-filled depressions in the granite outcrops of southwestern Australia was studied by Bayly (1982). This region is characterized by dry summers and wet winters and pools form in mid-May. The pools are generally shallow (most are less than 10cm deep), have a pH of 6.0 or less and have a conductivity of less than 180$\mu$S/cm. Some of the pools have a pH of less than 4.0, probably the result of plant growth on the pool bottoms. Daytime water temperatures range between 10.7 and 17.2$^{o}$C. The fauna is dominated by microcrustaceans, particularly Cladocera, Ostracoda and Copepoda (of which the cladoceran *Neothrix armata* is endemic) and Acari (especially oribatids) and Chironomidae. Other groups represented are Turbellaria (flatworms), Anostraca, Conchostraca, Hemiptera, Coleoptera, Nematoda and Ceratopogonidae. The last group is represented by *Dasyhelea*, the same genus found in rock pools in Africa, but apart from this the composition of the faunas in not very similar between the two continents.

Some succession of species is evident in these habitats. Food input in some pools is derived from the breakdown of lichens and mosses growing on the granite surfaces while, in others, marsupial dung is the major source of energy.

### *4.2.8 Antarctic melt-water streams*

Melt-water streams are common around the margins of Antarctica and may vary in length from a few to tens of kilometres; some flow from ice sheets to the sea while others flow inland to lakes (Vincent and Howard-Williams, 1986). In the McMurdo Sound region of southern Victoria Island ($78^{o}$S; $165^{o}$E) the streams are characterized by highly variable flow patterns based on diel, seasonal and annual periodicities. Typically, streamflow begins in mid November-early December and continues into January, but there is variation between years. By late January-early February, the streams freeze solid. Smaller streams are particularly subject to diel changes in flow, with discharge changing as much as two hundred-fold according to whether the source-glacier face is exposed to direct sunlight or is in shadow (e.g. range $0.006m^3$/s, in shadow to $0.1m^3$/s, in direct sunlight; as measured in the Whangamata Stream originating at the Commonwealth Glacier; Vincent and Howard-Williams, 1986).

Associated with these dramatic changes in discharge are wide fluctuations in water temperature and dissolved nutrients. Water temperatures are invariably low but vary both from day to day (e.g. from 3.0 to $5.5^{o}$C) and diurnally (e.g. from $0^{o}$C at 2:00 a.m. to $6.0^{o}$C at 6:00 p.m.). Suspended solids tend to be positively correlated with discharge, while dissolved nutrients (e.g. nitrates and soluble reactive phosphorus) are highest when the streams start to flow but their concentrations decline thereafter.

The epilithic community of the rock surfaces is simple, compared with those of temperate streams, and is dominated by filamentous algae (especially *Nostoc, Phormidium* and *Oscillatoria*) but also contains bacteria, fungi and microherbivores such as protozoans, rotifers, tardigrades and nematode worms. Larger herbivores, such as aquatic insects and crustaceans are completely absent from this region of Antarctica, although three species of Collembola (springtails) and three species of mites are known to occur in riparian mosses (Vincent and Howard-Williams, 1986).

### *4.2.9 Temporary saline ponds and lakes*

In addition to having to contend with water loss from their habitat, the biota of temporary saline waters has to deal with change in ionic proportions. As the water evaporates, the ions become more concentrated, thus, perhaps, adding an osmoregulatory stress-factor to those environmental forces already acting on the biota. In saline ponds, the ions most commonly present are sodium and chloride but magnesium, calcium and sulphate ions may also occur. Salinity may vary, over one year, from $<50^{o}$/oo to $>300^{o}$/oo and the pattern of variation may change between years (Figure 4.10).

As well as acting through changes in osmoregulation, fluctua-

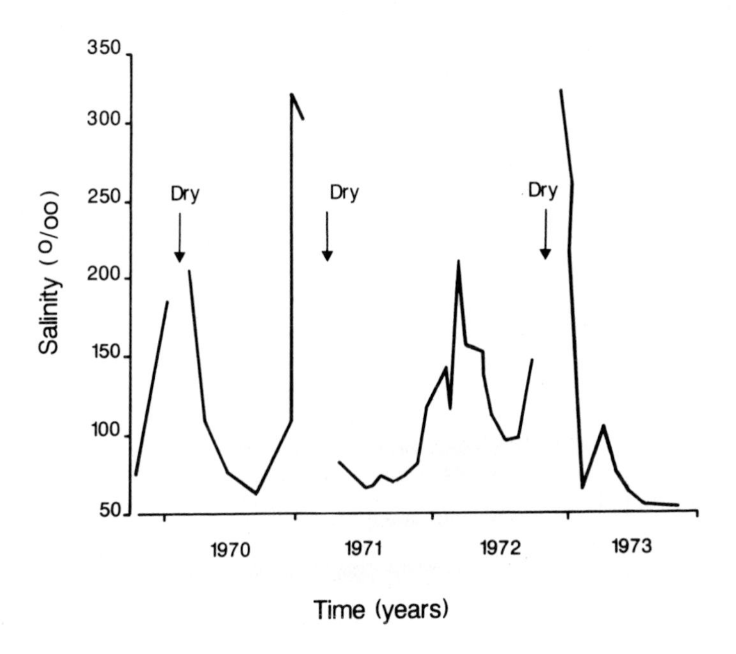


Figure 4.10: Salinity of Lake Eurack, Victoria, Australia (redrawn from Williams and Buckney, 1976).

tions in salinity may influence organismal respiration as less oxygen can be dissolved in saline water than in fresh water. Other factors, such as turbidity and high seasonal and diel fluctuations in water temperature, experienced by freshwater temporary pond floras and faunas are experienced also by the inhabitants of saline temporary ponds.

It was believed, until recently, that the diversity of organisms in temporary saline habitats was low and that those species present were more or less cosmopolitan in distribution. This is now thought not to be the case (Williams, 1981b) as the following groups are now known to be represented, although frequently not in the same habitat or geographical area: ciliates, foraminiferans, spirochaetes, gastropods, oligochaetes, anostracans, copepods, cladocerans, chironomids, ephydrids, ceratopogonids, culicids, stratiomyids and birds. Williams (1985) pointed out the notable absence of obligate temporary water forms such as the notostracans and conchostracans, and also the general lack of frogs and fish.

Faunal diversity tends to be negatively correlated with habitat salinity although the relationship is not clear-cut. In fact, within wide salinity boundaries, salinity may not be a prime determinant of species presence (Williams, 1984). Diversity also may be linked to habitat predictability, with lakes subject to predictable filling (e.g. Lake Eurack, Australia) having more species than unpredictably-filled ones (e.g. Lake Eyre, Australia). Although some seasonal succession in the species is evident, it does not appear to be as well defined as in freshwater temporary ponds (Figure 4.11).

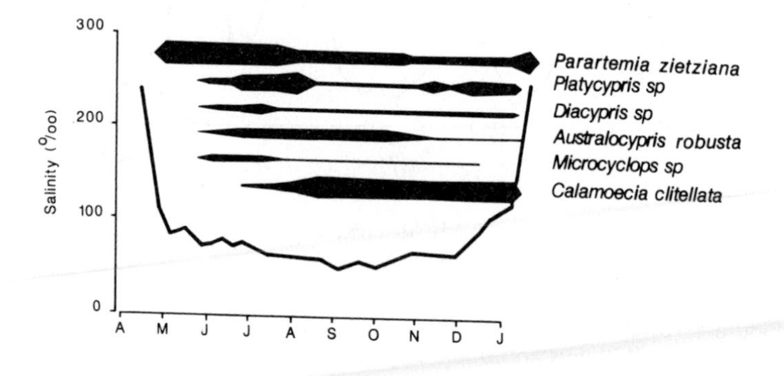


Figure 4.11: Selected examples of the seasonal distribution of species in Lake Eurack, Victoria, Australia (redrawn from Geddes, 1976).

Little is known of the micro- and macroflora of temporary saline ponds. The bacteria *Halobacterium* and *Halococcus* have been found in highly saline waters. Above 100‰, the blue-green alga *Aphanothece halophytica* occurs and its distribution appears cosmopolitan. Below 100‰, other blue-greens appear, e.g. *Oscillatoria, Phormidium, Microleus* and *Spirulina*. The green alga *Dunaliella* occurs at high salinities but others occur only at moderate salinities. Few submergent macrophytes tolerate much salinity but *Ruppia* (Potamogetonaceae) tolerates salt well. *Scirpus maritimus* (bulrush) is an important emergent halophyte in North America (Williams, 1985). Salinity maxima at which various species have been recorded in the field are summarized in Table 4.4. It is clear from this that adaptation to high salt levels in combination with resistance to desiccation is something that quite a wide variety of organisms have achieved.

Table 4.4: Maximum salinities at which various species have been collected from temporary salt lakes (from Williams, 1985).

| Major group | Taxon | Salinity ($^0/_{00}$) |
|---|---|---|
| Bacteria | *Halococcus* sp. | 350 |
| | *Halobacterium* spp. | 350 |
| | *Ectothiorhodospira* spp. | 340 |
| Cyanobacteria | *Spirulina* sp. | 100 |
| | *Aphanothece halophytica* | 350 |
| | *Oscillatoria limnetica* | 250 |
| Algae | *Dunaliella salina* | 350 |
| Macrophytes | *Ruppia* spp. | 230* |
| Anostraca | *Artemia salina* | 330 |
| | *Parartemia zietziana* | 353* |
| Ostracoda | *Diacypris whitei* | 180* |
| | *Limnocythere staplini* | 205 |
| Copepoda | *Calamoecia salina* | 131* |
| | *C. clitellata* | 113* |
| Isopoda | *Haloniscus searlei* | 192* |
| Mollusca | *Coxiella striata* | 112* |
| Insecta | *Ephydra* sp. | 222 |
| | *Culicoides variipennis* | 220 |
| | *Tanytarsus barbitarsis* | 95* |
| Fish | *Taeniomembras microstomum* | 70* |

*Australian endemics

Problems of living in waterbodies subject to varying levels of salinity are chiefly ones of osmoregulation. Williams (1985) identified two mechanisms by which organisms cope with this problem: osmoregulation and osmoconformity, the former is typical of animals in saline habitats, the latter is found in all plants and in some animals. In osmoconformers, high internal pressures are maintained by the accumulation of organic or inorganic ion osmolytes. In the Cyanobacteria, sucrose, betaine, glutamate and glucosylglycerol have been identified as osmolytes (Borowitzka, 1981). In the Halobacteria, proteins within the cytoplasm undergo wide amino acid substitution and this enables cytosolic proteins to function under high concentrations of inorganic solutes - in this case potassium (Kushner, 1978). In many halophytic macrophytes, the osmolyte is proline.

The majority of animals living in temporary saline ponds tend to osmoregulate by taking in the water and then excreting unwanted ions. This is typically the method seen in *Artemia* (Anostraca), *Ephydra* (Diptera) and *Haloniscus* (Isopoda). However, some animals appear to osmoconform although really they may be osmoregulating at the cellular level, that is, osmoregulation may be a function of cellular activity, or cells must have a high tolerance of salt (Williams, 1985). The copepod *Calamoecia salina* of Australia is an osmoconformer (Bayly, 1969). Some animals seem to be able to switch so that, for example, the snail *Coxiella striata* osmoregulates weakly at low salinities but osmoconforms at high salinities (Mellor, 1979).

In some osmoconformers (e.g. the alga *Dunaliella*, Halobacteria and Cyanobacteria) the osmolytes have not only an osmoregulatory function but they also appear to reduce the effect of cold and heat on enzyme systems (Borowitzka, 1981). The latter is particularly useful in temporary saline ponds in arid regions where water temperatures frequently exceed 30$^{o}$C.

## 4.3 Other inhabitants of temporary waters:

Thus far, mostly the invertebrate fauna has been mentioned. Unfortunately, the dearth of studies on the microflora, macrophytes, protozoans and vertebrates of temporary waters makes it impossible to discuss these groups in any other context than on the local scale. For convenience, any information available concerning adaptations of these groups to life in temporary waters is included here rather than in the later chapter devoted to that topic.

### *Bacteria*

Felton *et al.* (1967) found nine physiological groups of bacteria in a temporary pond in Louisiana, U.S.A. Sulphur-, ammonia-, and nitrite-oxidizing autotrophs were absent. They concluded that bacteria did not make a major contribution to the pond ecosystem as primary producers. Counts of aerobic nitrogen-fixers and urea-using forms ranged from $10^6$ to $10^8$/g of bottom mud suggesting a significant role in the nitrogen cycle in the pond. Bacteria capable of decomposing cellulose were found both in the water column and in the mud. Mud samples taken in the spring contained fewer cellulose-decomposers than samples taken in the autumn when, presumably, terrestrial plant material growing in the basin during the summer dry period would be starting to die and decay.

Heterotrophs (those species requiring a supply of organic material from their environment) were the largest physiological group found in the pond, with anaerobes being more abundant than aerobes. Counts of both types decreased as the pond dried up but later increased after the basin was dry. Heat-shock experiments revealed that about 1.5% of the aerobic heterotroph population consisted of spore forms (heat resistant). The authors concluded that the bacteria functioned as decomposers and transformers in the nitrogen, carbon and energy cycles of the pond as well as acting as a source of nutrients and as a primary source of food for protozoans and plankton.

*Fungi*

In a study of a temporary vernal pond in southern Ontario, Barlocher *et al.* (1978) found the fungal flora on bottom detritus to be very different from that found in nearby permanent ponds. Terrestrial fungi predominated during the waterless period and a seasonal succession was evident.

*Algae*

Sheath and Hellebust (1978) studied the various algal communities in an arctic tundra pond and also reviewed the literature on these very abundant habitats. Tundra ponds develop where vegetation has been injured resulting in an increased thawing of permafrost, where there is a loss of volume as ice is melted, and where subsidence of sediments occurs. These ponds hold water during the brief summer (June-August) but freeze solid under 2m of ice each winter.

The plankton community shows two peaks of biomass and primary production during the short ice-free season. Only one peak in biomass has been recorded for the periphyton community but primary productivity may show one or two peaks.

Antarctic streams have been shown to support rich epilithic communities of high biomass (>20$\mu$g chlorophyll $a$/cm$^2$, or >20mg

C/cm$^2$), yet production rates are low (Vincent and Howard-Williams, 1986). Nutrient supply and light are not limiting factors in these streams but water temperatures seldom rise above 5.0°C and, more often, lie between 0 and 2.0°C. Metabolism in these algae thus occurs at a low rate and the high biomass observed represents accumulation over several seasons of growth (as theoretical turnover times were of the order of several hundred days yet each annual growing season lasts less than 80 days). Despite this, the overwintering community retains a high metabolic capacity and responds rapidly to hydration at the beginning of summer. Figure 4.12 shows that photosynthetic capacity rises as a log function of time over the first six hours, and then at a faster rate over the subsequent two days. Thus simple rehydration allows immediate resumption of some photosynthesis but full recovery necessitates longer-term biosynthesis and repair. Vincent and Howard-Williams (1986) likened this resurrection to the response of desert plants in warmer regions, each community inhabiting a seasonally arid environment. They pointed out, however, that in addition to loss of water, Antarctic epilithon must contend with continuous darkness in winter and a harsh freeze-thaw cycle. Physiological resilience to freezing must therefore be an essential property of cyanophytes in the Antarctic.

In temperate ponds, algae survive prolonged exposure to drying as modified vegetative cells with thickened walls, mucilage sheaths and accumulation of oils in the cells. The ability to resist exposure by such means is the major factor controlling zonation of algae at pond margins.

Species of algae in temporary ponds and streams are opportunists. Many pass through predictable life cycle phases with maximum zygospore germination occurring when water levels are highest. *Vaucheria*, a typical temporary pool alga, survives drought as the thickwalled zygotes discussed earlier but it also has a "back-up" system. This is to form hypnospores in response to rapid desiccation. These are specialized structures which release amoeboid cells capable of movement to areas where water is more abundant and where they give rise to new filaments (Sands, 1981).

### *Macrophytes*

As previously indicated, most temporary ponds represent the penultimate stage of a sere, the climax of which is terrestrial. In a study of plant succession in temporary ponds in Oregon, U.S.A., Lip-

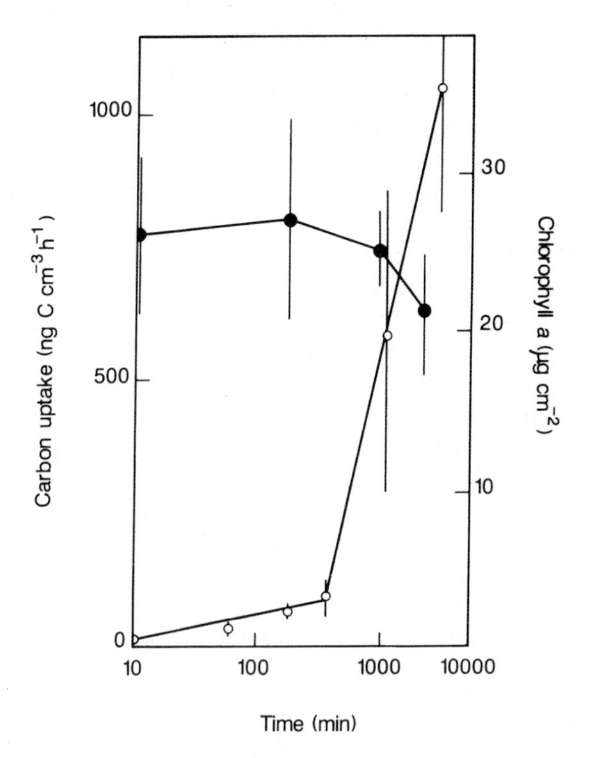


Figure 4.12: Photosynthetic recovery by the *Phormidium* epilithon as a log function of hydration time (closed circles = chlorophyll *a* content: open circles = photosynthesis; each point is the mean of three samples ± 2 standard errors; redrawn from Vincent and Howard-Williams, 1986).

pert and Jameson (1964) found the species present were characteristically those found in wet places, for example, the cat-tail (*Typha latifolia*) and the spike rush (*Eleocharis palustris*). In northern Germany, Caspers and Heckman (1981) found *T. latifolia* and the grass *Glycera maxima* in large numbers in stage 5 (the final stage) of their ditches. They stated that these species are eventually responsible for the total disappearance of the aquatic habitat. The margins of temporary waters are very susceptible to invasion by terrestrial species, the roots of which may proliferate through the basin substrate. Litter from all species provides a rich substrate for microorganisms and subsequently a source of food for aquatic invertebrates. In certain localities, free-floating annual macrophytes such as *Azolla* may contribute significantly to the productivity of temporary ponds.

In a study of the microdistribution of macrophytes of vernal pools in southern California, Zedler (1981) related water level to a succession of species groups each with characteristic germination times, and flowering and seed maturation stages. Species were assigned to weighted average water duration classes (WADC), calculated from duration - frequency data, ranging from 1 to 16. Species falling into classes 1 to 4 are those rarely found within the inundated areas of pools but which grow commonly near the margins (e.g. *Bromus* spp., bromegrass; and *Erodium* spp., filaree). Species with WADC values of between 4.5 and 7.0 (e.g. *Juncus bufonius*, toad rush; and *Agrostis microphylla*, bentgrass) possess a physiology capable of tolerating inundation but fall short of being true marsh or aquatic plants. Species in classes 7.0 to 9.5 tolerate water-cover for long periods, but do not thrive under long duration - they are able to germinate and grow underwater but require a considerable period out of water in order to mature (e.g. *Callitriche marginata*, water-starwort; and *Anagallis minimus*, false pimpernel). Characteristic temporary pool species fall into the class range 9.5-11.5. These are species possessing the morphological and physiological plasticity that allows them to withstand prolonged submergence (e.g. *Downingia cuspidata*; *Pogogyne abramsii*, pogogyne; and *Eryngium aristulatum*, marsh eryngo). All three species produce submerged leaves that are very different from the emergent foliage of the mature plant. The final group of plants (classes 11.5-16) contains species that are almost true aquatics, yet they can withstand some dry period (e.g. *Pilularia americana*, pillwort; *Callitriche longipedunculata*, water-starwort; and *Lilaea scilloides*, flowering quillwort).

On the basis of this study, Zedler proposed several hypotheses. The first was that the distribution of standing water in time and space is the single most important factor influencing temporary pool macrophytes. Physical stress of inundation is the primary cause of the distinctive assemblages of species but pattern of soil moisture may be important also. The second proposal was that local extinctions of temporary pond species are rare despite considerable year-to-year variations in rainfall and habitat availability. Several characteristics of temporary pool species lessen the probability of extinction, namely small minimum plant size; small seed size and thus many seeds/unit biomass; and high vegetative and reproductive plasticity typical of annuals. Holland and Jain (1981), however, in a survey of over 250 vernal pools in central California, showed that species composition varied significantly between years and that species richness varied nearly two-fold over sites, apparently in response to regional differences in rainfall among sites and edaphic conditions, and was less in a drought year. A third hypothesis was that despite the fundamental influence of inundation on community structure, competition is also a factor. Most temporary pool species are restricted to the upper (drier) end of the water duration-elevation gradient by competition from species which are better at exploiting drier conditions. At the lower (wetter) end, however, morphological and physiological tolerance to inundation is probably more influential. Holland and Jain showed that several grassland species grew in vernal pools during a year of severe drought but they were excluded in years of average rainfall. Further, several taxa characteristic of the vernal pool flora in this region were not evident during the drought though they were common and widespread in subsequent years. They showed that the number of species specialized for a given pool depth is proportional to the relative area in a pool at that depth; thus more species are adapted to the margins than to the centre. They concluded that species richness is determined chiefly by certain physiographic niche properties whereas competitive niche-partitioning factors influence congeneric sympatry and zonation within a pool.

*Protozoa*

Fenchel (1975) found densities of around $10^6/m^2$ and biomass of 20-40mg/$m^2$ for ciliates in an arctic tundra pond between June and August. Some fed on algae or bacteria, while others were carnivores feeding on zooflagellates and other ciliates.

*Fishes*

Apart from highly specialized forms such as lungfishes which can aestivate in the bottom mud during the dry phase, fishes tend to be absent from temporary ponds. Temporary streams, on the other hand, may support fish populations of considerable size and diversity. This is the result of either some streams becoming intermittently connected to permanent waters from which fishes can migrate, or fishes surviving in permanent pools left in some drying streambeds. Studies of the ecology of fishes in temporary streams are, however, few in number, possibly because although large rivers do occasionally dry up, particularly in the tropics - intermittency, in temperate regions, is more a characteristic of smaller bodies of water which usually support populations of fishes of little economic or recreative importance.

In the Kalahari Desert, three species of fish can survive. *Clarias gariepinus* is a catfish which possesses suprabranchial respiratory organs. The other two species are both cichilds, *Tilapia sparrmanni* and *Hemihaplochromis philander*, all three species can survive in very little water as stunted individuals (Cole, 1968).

In North America, out of a total of 50 species of fish found by Williams and Coad (1979) in the Grand River watershed, Ontario, only 12 were collected from three local temporary streams. Species in the latter habitats were largely members of the families Cyprinidae (minnows) and Percidae (perches), with one species from each of the Catostomidae (suckers) and Gasterosteidae (sticklebacks). The temporary streams entirely lacked catfishes, sunfishes and salmonids. The brook stickleback *Culaea inconstans* and the cyprinids *Pimphales notatus* and *P. promelas* show physiological tolerance in that they survive poor water quality, high temperatures and crowding in summer pools. *Catostomus commersoni* (white sucker) and *Semotilus atromaculatus* (creek chub) on the other hand, move into the temporary streams, from larger rivers, to spawn and then leave. The main advantages to fish species colonizing temporary streams appear to be plentiful food, earlier spring breeding (as the water is often warmer than in adjacent permanent streams) and reduced predation by large fishes.

A problem faced by fishes moving into temporary streams is that they may become stranded and die if the pools dry up completely. The longfin dace, *Agosia chrysogaster*, possesses behavioural adaptations that contribute to its success as the only species to consistently use intermittent streams in the Sonoran Desert of Arizona. These streams dry up to form pools separated by lengths of dry streambed. The fishes position themselves in the current and this minimizes the chances of them becoming stranded by falling water levels. Avoidance of the stream edges and shallows reduces predation from birds and mammals. The species quickly invades new habitats during wet periods when flow is continuous and is capable of existing for at least 14 days in areas where there is no free water, provided that there is moisture beneath mats of algae (Minckley and Barber, 1979; Bushdosh, 1982).

Other aspects of fishes in temporary waters will be discussed in Chapter 8.

### *Amphibians*

Frogs and salamanders are common inhabitants of temporary ponds and cosmopolitan genera include *Rana, Hyla,* and *Ambystoma.* Clearly, though, the aquatic larvae of these animals are severely threatened by any untimely onset of the dry phase of the habitat. Wilbur and Collins (1973) have suggested that an endocrine-controlled, metabolic-feedback mechanism exists in temporary pond species. Should the rate of larval growth be slow, metamorphosis to the adult stage is initiated once a certain minimal larval size is attained. Although the resulting small adult may face disadvantages in the terrestrial environment, these are less than those facing the larva if the pond dries up prematurely. If, however, the larval body size is small but its rate of growth is fast, metamorphosis is delayed so as to maximize the animal's growth potential in the pond. Control of metamorphosis is thus related to the stability of the habitat and species with a fixed size for metamorphosis are therefore excluded from temporary waters.

In Western Australia, all the species of *Heleioporus* breed in the winter and lay their eggs in a frothy mass in a burrow dug by the male. The site chosen is always one which will be later flooded by heavy winter rains. Species of *Pseudophryne* have similar egg-laying habits. As the rain raises the level of the water table in the burrow, the larvae break free of the egg mass and develop to metamorphosis often in no more than 0.5 litres of water. In the two western species of *Pseudophryne,* larval development takes slightly longer than 40 days. The aquatic phase of their ponds seldom lasts more than 50 days so potentially there is little leeway. This is offset, however, by the ability of the eggs to complete embryonic development (6-8 days) in the absence of free water, so that the larvae are ready to hatch as soon as the rains come. In the event of a delay in rainfall, hatching can be postponed for several weeks (Harrison, 1922; Main *et al.,* 1959).

Adults of the five species of Australian *Heleioporus,* together with those of the genus *Neobatrachus,* can withstand a drop in the water content of their bodies of up to 45% of their body weight.

Rehydration rates in species of *Neobatrachus* vary according to the severity of water loss in their particular habitats. For example, species that live in the arid interior rehydrate faster than those from the wet coastal regions of the southwest. In contrast, species of *Heleioporus* show no difference in rehydration rates of species occurring across a wide spectrum of aridity. It is thought that because all species of *Heleioporus* are superior burrowers, selective pressures for increasing the speed of rehydration may not operate (Bentley, 1966).

Burrowing seems to be a common method of surviving droughts in amphibians, as even the ability to get just a few centimetres below ground level places the animal away from the drying effects of sun and wind and into a more moist environment. Even in deserts, moisture from past rains can remain trapped for years in sand at depths of only 20-30cm (Bagnold, 1954).

*Scaphiopus couchi*, the spade-foot toad of California aestivates in its burrow surrounded by a layer of dried, skin-like material. This may help to limit evaporation of moisture from its body in much the same way as the cocoon-like covering of the African lungfish. Another species, *Scaphiopus hammondi* lines its burrow with a gelatinous substance which presumably slows down water loss. In this genus, aggregations of tadpoles have been found shortly before metamorphosis and subsequent emergence from temporary ponds (Bragg, 1944). It has been suggested that these dense aggregations may conserve water, as the combined rapid beating of many tadpole tails tends to deepen that part of the pond basin and water from shallower parts of the pond will drain into it.

That amphibians, particularly frogs and toads, can successfully contend with the intermittent availability of water (though most people think of them as being associated with cool, moist environments) is evidenced by phenomena such as those that occur in the western deserts of Australia. Here, after rain, the number of frogs emerging from burrows is so large as to interfere with rail transportation, as thousands of frogs are crushed as they attempt to cross railway lines thus making traction impossible (Bentley, 1966).

Mayhew (1968) summarized the general adaptations of amphibians found in dry areas as follows:

- no definite breeding season,
- use of temporary waterbodies for reproduction,
- breeding behaviour initialized by rainfall,
- loud voice in male attracts both females and other males, resulting in rapid congregation of breeding animals,

- rapid development of eggs and larvae,
- omnivorous feeding habits of tadpoles,
- production of inhibiting substances by tadpoles which influence the growth of other tadpoles,
- high tolerance of heat by tadpoles,
- adults have metatarsal "spades" for digging,
- ability to withstand considerable dehydration, compared with other anurans,
- nocturnal activity.

A particularly noteworthy observation on amphibian populations in temporary waters is that there exists a great deal of variation in reproductive characteristics (especially in egg and clutch size) between individuals. Kaplan (1981) has examined this variation in the light of current theory on "adaptive coin-flipping" or natural selection for random individual variation. Given two genotypes with equal mean fitnesses, the genotype with less variance in fitness will eventually outcompete the genotype with more variable fitness (Felsenstein, 1976). Kaplan argues that natural selection for random individual variation has been overlooked because, in general, neo-Darwinian theory is a theory of genes and not a theory of development and, consequently, population geneticists do not take into consideration influences of the environment on development. A well-buffered phenotype may be advantageous in many cases, but a less well-buffered developmental system might also be of advantage to an individual, particularly in environments that are temporally variable. This is supported by the observations made on variation in size for metamorphosis discussed earlier.

*Other vertebrates*

Ehrenfeld (1970) found that alligator holes (temporary ponds excavated by the reptiles) in the Florida Everglades served as collecting ponds and biological reservoirs for the surrounding aquatic life - both vertebrate and invertebrate - in the dry season. Rich growth of algae and higher aquatic plants were nourished by the reptiles' droppings and these, in turn, maintained a variety of animal life. At the end of the drought, the survivors moved out to colonize the glades anew. In Kenya, some waterholes are formed from the erosion of termitaria by wildlife such as rhinoceros and hartebeest rubbing against the mounds. When subsequently weathered below soil level, water collects in them and elephants, warthogs, buffalo and other animals use them as sources of drinking water, thus accelerating the drying-out process (Ayeni, 1977).

In many arid or semi-arid regions of the world, scattered water holes provide drinking water for animals that can move long distances. Most large mammals in areas such as the grasslands of Africa generally drink every one or two days, in hot weather, and are thus very dependent on finding waterbodies regularly. Only a few species of ungulates are capable of going without water for longer periods (e.g. camels), though there are a number of rodents (e.g. kangaroo rats, gerbils, pocket mice, jerboas) that seem capable of existing on water obtained from food alone.

In a study of the use of a vernal pool by small mammals in chaparral and coastal sage scrub communities in southern California, Winfield *et al.* (1981) found that it did not appear to be used heavily. This was despite the fact that it provided a potential source of food in the form of protracted growth of riparian vegetation and aquatic ground-dwelling insects. Of seven common species of mouse, rat and rabbit, only one, *Reithrodontomys megalotis* the western harvest mouse, had a higher estimated population at the pool than elsewhere, but these results were considered tentative.

In Africa, many species of wildlife move onto the floodplains of rivers during the dry season in search of grazing and prey. Some antelopes, such as the bush buck, *Tragelaphus*, and the lechwe, *Kobus*, migrate back and forth across swampy ground as the floods rise and fall. Their life cycles are aptly timed such that they drop their young as the floodwaters recede and new pasture is exposed (Welcomme, 1979). Hippopotamus are important transporters of nutrients through their habits of grazing on floodplains at night and depositing large quantities of nutrient-rich dung in water as they bathe during the day. The capybara, *Hydrochoerus*, of South America similarly inhabits floodplains where it feeds on grasses and aquatic plants; it is generally associated with permanently wet areas (Gonzales-Jimerez, 1977).

A variety of terrestrial reptiles may use temporary waters as sources of drinking water throughout the world. In addition to these, there are several families of reptiles that live in close association with both permanent and temporary waterbodies, for example crocodiles, monitors, turtles and iguanas.

In Australia, extensive breeding of waterfowl occurs in floodplain areas adjacent to rivers. In lightly wooded and treeless plains, vast areas of new waterfowl habitat are created when rivers overflow their banks. These shallow waters are soon colonized by huge numbers of grey teal, pink-eared ducks and shovelers, and by lesser numbers of black duck and white-eyed duck (Frith, 1959). Piscivorous bird species also have life cycles closely linked to floods such that, in Africa, fledgelings are produced just when small fish appear in flood-plain pools. There is also heavy predation as the floodwaters recede and fishes become stranded in temporary pools and channels (Bonetto, 1975).

## 4.4 Comparison of permanent and temporary water faunas:

Wiggins *et al.* (1980) maintain that temporary waters represent discrete types of freshwater habitats in which the dry phase imposes such rigorous environmental conditions that only a limited number of species can survive in them. Nevertheless, as we saw earlier, species from most of the major groups occurring in freshwater have succeeded in doing so. This reflects the richness of resources in these waters and the powerful selection forces at work in all these animal groups. Water in an aquatic environment is surely a fundamentally necessary factor. However, a large proportion of temporary water species are insects or mites, groups which evolved on land, and the problems of adapting to water loss may, therefore, not be as difficult as first thought; diversity in temporary waters is often high. There are, perhaps, no less rigorous factors controlling the community composition of permanent waterbodies.

Comparison of community composition between permanent and temporary waterbodies reveals relatively little overlap. Two examples from Canada are given in Table 4.5. In example A, the temporary stream is actually a tributary of the permanent stream, potentially allowing easy access between the two. However, only 11.3% of the taxa overlap. In example B, the two streams, although not connected, are only 200m apart yet the similarity index is only 13.3%. In contrast, in some parts of Australia, much of the "typical" stream fauna present in nearby permanent streams can be found successfully reproducing in temporary streams (Boulton and Suter, 1986).

Dance and Hynes (1979) found that two tributaries of the same Ontario river, one permanent, the other temporary, both had rich midge faunas. At the species level, however, some of the Chironominae and Tanypodinae occurred only in the intermittent branch, while the Diamesinae and some Orthocladiinae only occurred in the permanent one. In a study of calanoid copepods from permanent and

Table 4.5: Similarity in taxa between nearctic temporary and permanent streams.

| Example A. | Temporary stream | | Permanent stream |
|---|---|---|---|
| Number of taxa | 33 | | 46 |
| Number of taxa in common | | 8 | |
| Similarity index | | 11.3% | |
| **Example B.** | | | |
| Number of taxa | 23 | | 28 |
| Number of taxa in common | | 6 | |
| Similarity index | | 13.3% | |

temporary ponds in Arizona, Cole (1966) collected six species in abundance. *Diaptomus albuquerquensis*, *D. siciloides* and *D. clavipes* were common to both types of pond. *D. nudus* occurred only in one permanent pond and *D. novamexicanus* and *D. sanguineus* were restricted to temporary ponds. Comparative measurements of each of the three species common to the two pond types revealed that cephalothorax lengths and clutch sizes were greater in the temporary pond populations than in the permanent pond populations. Congeneric occurrence of adjacent-sized and similar-sized species was common in the temporary ponds. Cole cited reasons for these observations as being either genetic or the result of superior food supply in the temporary ponds. Predation on large individuals might select for smaller individuals in permanent ponds.

Otto (1976) compared the biology of three species of cased caddisflies from southern Sweden. *Limnephilus rhombicus* occurred in a pond, *Glyphotaelius pellucidus* occurred in a temporary pool and *Potomophylax cingulatus* was collected from a small stream. *L. rhombicus* was found to be poor at coping with water level fluctuations and with fast water. *G. pellucidus* was inferior to the other species in interspecific interactions but endured low oxygen concentrations best. *P. cingulatus* was confined to areas of high oxygen but was superior in resisting the force of water currents.

Even intraspecific (within species) differences are noticeable between the two habitat types. Way *et al.* (1980), for example, found that a population of the clam *Musculium partumeium* in a temporary pond had a univoltine life cycle spanning 13 months. In contrast, a population from a permanent pond went through two generations in the same period of time. Springtime recruits to the population had high C:N ratios. This suggests that they have a high content of fats and/or carbohydrates in their tissues which sustains them during the dormant phase which begins at birth. The life history characteristics of *M. partumeium* do not appear to fit exactly any of the current theories on life history tactics - for example, that temporary pond populations tend to show more features of K-selected species rather than *r*-selected species (see Chapter 6) (Burky *et al.*, 1985). Clams in the permanent pond appeared to mimic the life cycle pattern of the temporary pond population in that they went through a period of "dormancy" during the summer which coincided with the dry period of the temporary pond. It has been suggested that this species actually evolved in temporary habitats and thus the summer "dormancy" noted in permanent pond populations may be a genetic legacy.

The brine shrimp *Artemia salina* represents a complex of sibling and semi-species that inhabit permanent lakes and temporary ponds which are salty. In Mono Lake, a permanent habitat in California, the *Artemia* population is highly specialized to live in a permanent lake. The cues for hatching and cyst formation seen in temporary ponds in Nevada (desiccation, increased salinity, high temperature, rehydration) are not present in Mono Lake. Instead, the Mono Lake shrimps have evolved a cold, preincubation hatching requirement, one more suited to a stable temperate lake. In addition, the cysts of the Mono Lake shrimps sink to the lake bottom and have a poor resistance to desiccation, properties related perhaps to a less crucial need for dispersal in a permanent environment. Again, the ability to survive passage through the digestive system of birds (thought to be an important mechanism of dispersal in *Artemia*) is not evident in the Mono Lake population. *Artemia* in Mono Lake have the fewest number of generations per year (two) known and this may support predictions from life history theory that species in stable environments tend to have fewer generations per year together with slower development of individuals (Dana, 1981).

So far, we have treated permanent and temporary freshwaters as separate entities. It is important to remember, though, that a continuum exists between the two. Caspers and Heckman (1981) in their study of orchard drainage ditches adjacent to the Elbe estuary found a definite seral progression towards a terrestrial climax. Distinct sets of physical and chemical conditions were evident along this gradient, together with several typical species aggregations appropriate to these conditions. In central Saskatchewan, Driver (1977) noted that midge diversity in small prairie ponds was dependent on the stage of development of the plant community along a moisture gradient. Generally, chironomid diversity increased gradually from temporary ponds to permanent ponds. From this, he suggested an interesting potential application - that the relative permanency of these ponds could be assessed by examining the chironomid faunas. This may be useful in situations where more obvious temporary water indicators, such as the Branchiopoda, are, perhaps for reasons of zoogeography, absent.

Figure 4.13 summarizes the changes in community composition which we would expect to see along the continuum. In ponds and streams that are dry for only a few weeks each year (left hand side of the graph) the community will consist largely of facultative species,

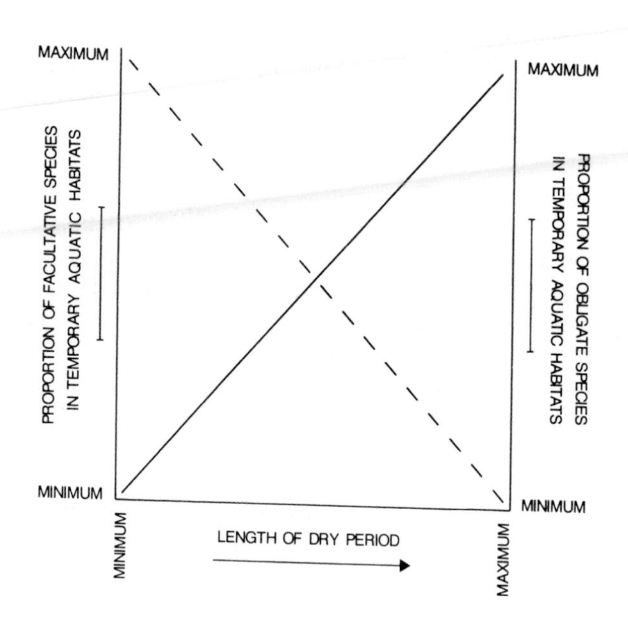


Figure 4.13: Changes in community composition along the temporary-permanent aquatic habitat continuum.

that is those found primarily in permanent waterbodies but which, because of wide physiological tolerance or flexibility of life cycles, are able to survive for short periods out of water. At the other end of the scale, in temporary waters that dry up for many months in a year, or even for periods of several years (as for example do some of the endorheic lakes in Africa and Australia) the community will be comprised entirely of obligate species which are highly adapted and restricted to temporary waters.

## 4.5 Community dynamics and trophic relationships:

In a study of the nutrient budget of a temporary pond in Massachusetts, Cole and Fisher (1979) found that the concentration of selected nutrients increased during the drying phase. Nitrate, phosphate and sulphate were retained in the pond during summer, while dissolved organic matter was released from the system. Primary production of macrophytes (primarily *Chara* sp.) was the largest source of organic carbon in the system (total production from all sources was estimated at 2140kcal/m$^2$/yr), while respiration of the heterotrophic community accounted for about 84% of the organic carbon loss (total estimated at 2669kcal/m$^2$/yr, with benthic animals accounting for almost seven times the metabolism of the plankton). The authors concluded that, during the summer months, small ponds are very open systems, in terms of nutrients, but are rather closed energetically. In comparison, Lewis and Gerking (1979) obtained a primary production rate of 19.1kcal/m$^2$/day and a community respiration rate of 15.48kcal/m$^2$/day in June for an unpolluted section of intermittent desert stream in Arizona (6954kcal/m$^2$/yr and 5650kcal/m$^2$/yr, primary production and community respiration, respectively extrapolated from the June rates - however, as June was the month of maximum flow, the authors caution placing too much value on annual figures). The major primary producers in this stream were the moss *Leptodictyum* and the alga *Cladophora*.

These two examples show us that temporary water systems may in some cases be autotrophic ($P/R>1$ as in the Arizona stream) and in others, heterotrophic ($P/R<1$) as in the Massachusetts pond). Food webs may not necessarily be based on these criteria as, for example, Ameen and Iversen (1978) observed that although a temporary pool in a Danish forest was highly heterotrophic, the grazing chain appeared to be the most important energy pathway for macroinvertebrates, especially mosquito larvae. Although the proportion of detritus and fungi in the guts of *Aedes communis* and *A. cantans* increased during larval development, it was never more than 50%.

As well as herbivores and detritivores, temporary waters support a number of predatory species. As we have seen, only under special circumstances are these fishes or other vertebrates. For the most part they are invertebrates such as the flatworms, leeches, crayfishes, some of the Diptera (true flies), and especially the dragonflies, bugs and beetles. Typically the appearance of predators lags behind that of the prey as prey density has first to build up to a level at which the predator can sustain itself. Subsequently, prey populations often decline and disappear completely. At first glance, this may appear to be the result of predation, as in classical terrestrial examples of predator-prey interrelationships, but it may well be due to abiotic changes in the habitat or entry into a new, "less conspicuous", stage in the life cycle, e.g. ephippial eggs in the case of cladocerans

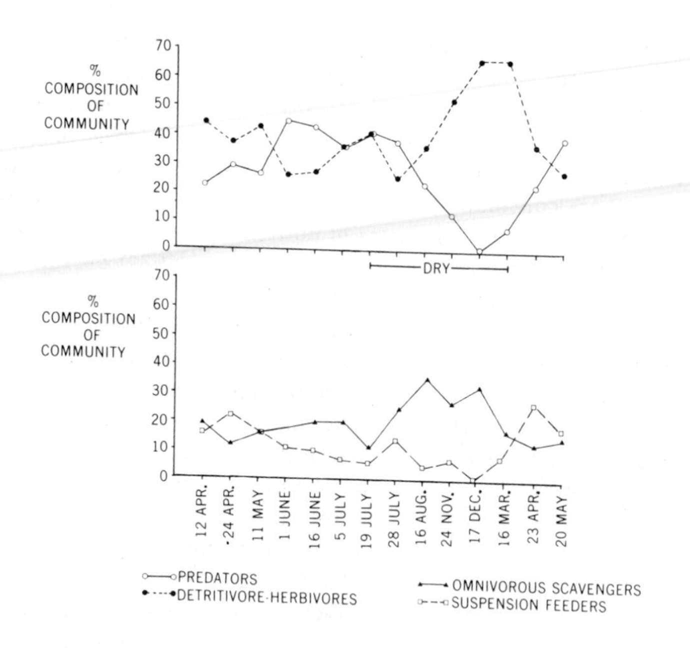


Figure 4.14: Seasonal changes in the percentage composition of the four main trophic categories in the Sunfish Pond community.

and emergence to winged adults in insects. Quite a few predators in temporary waters are casual invaders that fly in as adults to feed, and later fly out. We shall discuss this subject in more detail in the chapter on Colonization Patterns.

Seasonal changes in the percentage composition of the four main trophic categories in a temporary pond community are shown in Figure 4.14. The percentage of suspension-feeders generally declines throughout the aquatic phase, correlating with decreasing pond diameter and hence water volume. At spring thaw, the detritivore-herbivores predominate; many eating the decaying plant material left over from the terrestrial phase. Numbers of species in this category fluctuate throughout the aquatic phase and although they drop during the dry phase, they increase in importance in terms of percentage composition of the community - reflecting the movement of species into the drying basin to feed on decaying vegetation. The relative importance of omnivorous scavengers also increases at this time for similar reasons. The percentage of predatory taxa is relatively low at spring thaw but increases rapidly thereafter primarily due to the immigration of species capable of flight - particularly aquatic beetles, and remains high for most of the aquatic phase. Their importance declines shortly after dry-up.

## 4.6 Community structure - biotic versus abiotic control:

From a study of the changes in diversity of marine soft-sediment infaunal communities ranging across a gradient of decreasing environmental "stress" (shallow to deep water), Sanders (1968) proposed that where environmental conditions are severe and unpredictable, adaptations are primarily to the physical environment and such communities are physically controlled. Conversely, where environmental conditions are benign and predictable, adaptations are primarily directed at other species so as to optimize biological interactions. Menge and Sutherland (1976) criticized this view because of its failure to recognize (1) that animals are usually physiologically well-adapted to their environments and thus are rarely stressed by it, and (2) that all communities have interactions that occur between (predation) as well as within (competition) trophic levels. They argued that it is more appropriate to label a community as being "physically controlled" only if environmental episodes (catastrophies) are a primary and direct cause of distribution and abundance patterns; in such habitats, biological interactions are likely to be of lower intensity.

In temporary waters in which the dry period is cyclical and predictable, the communities will consist almost exclusively of obligate temporary water species, well adapted to environmental stress. Although such a community is ultimately controlled by physical constraints, it will be subject also to proximal control by biotic factors. Some of the specific temporary water types discussed at the beginning of this chapter support such communities in which species diversity is often very high. In temporary waters subject to unpredictable loss of water, the proximal factors controlling community structure are likely to be physical ones and species diversity should be predictably low. Temporary streams in Hong Kong seem to be good examples of these types of waters, as a combination of variable precipitation pattern, steep local relief and thin soils create unpredictably ephemeral habitats that support only one or two species of mosquitoes (Dudgeon, pers. comm.). Rain-filled rockpool communities in Africa show similar low diversity and are also controlled primarily by physical factors (McLachlan and Cantrell, 1980). Communities in permanent waters which experience isolated episodes of drought will also be subject, at these times, to control by physical factors although they may be controlled, normally, by biological ones.

Different types of temporary waters have different degrees of trophic complexity and the ways in which temporal heterogeneity of the habitat can affect community organization may increase with increasing trophic complexity. Menge and Sutherland (1976) identified some relationships for animal communities in general: trophic complexity is a function of temporal heterogeneity; the number of trophic levels in a community seems dependent, at least to some extent, on the rate and predictability of primary and secondary production, which is more than likely influenced by habitat stability and predictability. The, largely empirical, data on the trophic structures of temporary water communities suggest that there are fewer trophic levels as temporal heterogeneity increases. In other communities, this may be due to the elimination of more specialized consumers of high trophic status because the variety of resources becomes less predictably available. Another potentially important effect of temporal habitat heterogeneity on higher trophic species stems from the environment being unfavourable for certain periods (e.g. during drought in the case of an aquatic community). This reduces the time during which consumers can forage and may allow some prey species to avoid predation.

In communities in general, the often observed reductions in within-habitat species richness, along gradients of temporal heterogeneity, may be largely due to the increased incidence of competitive exclusion as trophic levels become lost or ineffective (Menge and Sutherland, 1976). However, the importance of interspecific competition for food resources in small seasonal ponds may be dependent on whether the resources are limited as, for example, may be the case in dytiscid beetles (Nilsson, 1986), and, where food resources are ephemeral, the probability of interspecific competition occurring is thought to be low thus allowing the coexistence of very similar beetle species (Price, 1984; Larson, 1985). Dytiscid diversity is less in more permanent waterbodies which may be a reflection of the more intense competition found there (Nilsson, 1986). The beetles themselves are known to sometimes exert considerable influences on some prey populations (e.g. *Acilius* larvae on *Daphnia*; Arts *et al.*, 1981) as has been shown for the effects of the predatory flatworm *Mesostoma ehrenbergi* on *Daphnia* in small ponds (Maly *et al.*, 1980), and for the effects of introduced fishes on the entire macroinvertebrate community of the pool stage of an intermittent stream (Smith, 1983).

# 5 SPECIAL ADAPTATIONS

## 5.1 Introduction:

As we have seen, in general, temporary waters exhibit amplitudes in both physical and chemical parameters that are much greater than those found in most permanent ponds and streams. The organisms that live in these types of habitat have therefore to be very well adapted to these conditions if they are to survive. Survival depends largely on exceptional physiological tolerance or effective immigration and emigration abilities. Transient species will be dealt with in the next chapter, while those that employ physiologically-oriented "strategies" will be dealt with here.

First, however, it will be useful to recap some of the environmental features of temporary waters that are most likely to affect the biota. In temporary streams, the period and range of waterflow are important factors affecting the flora and fauna as the current may range from torrential during spring floods to zero in the summer pools. The degree and rate of descent of the groundwater table and the permeability of the bed substrate are important also in so far as they control the formation and duration of the pool stage which invariably follows the cessation of flow. These pools allow the survival of certain species which would not normally be able to live in a lotic environment and, at the same time, invite colonization from many purely lentic forms. In tropical regions, the water temperature of temporary pools approaches the upper limit for biological processes, while in polar regions it seldom gets more than a few degrees above freezing. In temperate regions, water temperatures range from near zero to the low 30 s centigrade thus subjecting temperate species to temperatures near both the high and low thresholds for life. High temperatures in these pools encourage the rapid growth of algae which may supplement the food supply of the fauna but another consequence is depletion of oxygen dissolved in the water - especially at night. As we have already noted, along with increased levels of photosynthesis, dramatic changes in pH can be expected and these may affect the transport of materials across animal membranes.

The preceding discussion emphasizes the fact that adaptation is generally a multi-dimensional phenomenon, thus adaptation to a single environmental factor is not likely to ensure survival (Alderdice,

1972). The organisms living, successfully, in temporary aquatic habitats are more likely therefore to be adapted to deal with a number of habitat factor combinations. Studies of adaptation to such multivariable phenomena, particularly in temporary waters, are very rare (Williams, 1985).

But, perhaps by far the most influential environmental parameter affecting the biota is the loss of water during the dry season. For an aquatic organism, the imposition of what is in effect a terrestrial phase in its habitat must be a considerable obstacle to completing its life cycle. McLachlan and Cantrell (1980), in a study of survival "strategies" of fly larvae in tropical rain pools, found that the duration of the pool was important in determining which species was present. Large pools, lasting a few weeks after each rain, favoured the midge *Chironomus imicola* which has a larval life span of only 12 days. This species relies on egg-laying females to re-invade newly-filled pools. In situations where the pools were shorter-lived than this species' minimal life span some physiological mechanism would be needed in order to survive dry periods *in situ.* In practice though, these shorter-lived pools are inhabited by larvae of another midge, *Polypedilum vanderplanki* and larvae of the biting midge, *Dasyhelea thompsoni.* Larvae of *P. vanderplanki* are poor at invading newly-flooded pools, but are able to tolerate virtually a complete loss of body water (see later section) and are therefore able to survive drought in the dry mud. They are consequently always the first species there after refilling, a factor that gives them an advantage in very small pools. Larvae of *D. thompsoni* are not quite as good at surviving drought but are better at invading; they therefore occupy pools of intermediate size and probably inhabit a larger number of pools than either of the other two species.

The methods by which temporary water species survive the dry phase are varied but members of major taxonomic groups tend to employ similar stages in their life cycles. For example, mayflies, waterfleas and many midges survive as eggs; other microcrustaceans, amphipods and some midges survive as immature stages; while snails, some dragonflies and most beetles and true bugs (Hemiptera) survive as adults. We shall return to this topic in more detail later. Figure 5.1 is a pictorial summary of where these various aquatic groups survive the dry phase in an idealized temporary stream or pond basin. It shows eight basic means by which the fauna can cope with a summer drought, ranging from survival as adults on the wing to those species that burrow into the substrate or make use of the remaining

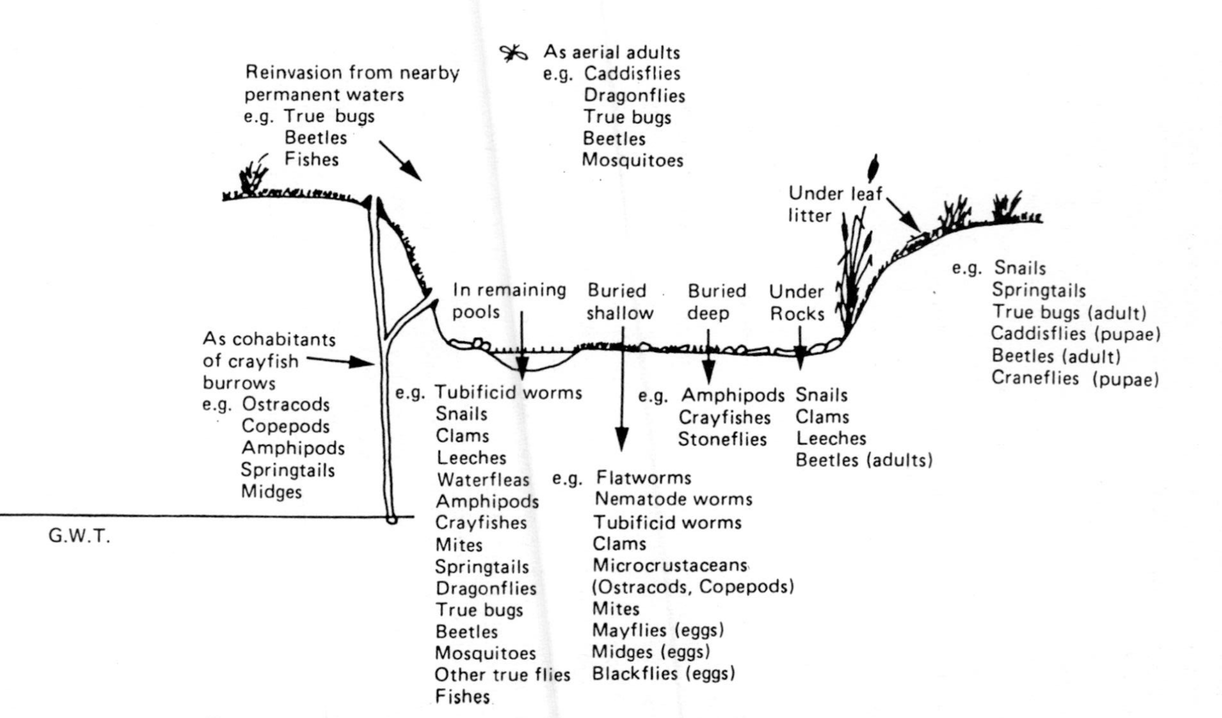


Figure 5.1: Summary of the drought-resistant methods employed by aquatic organisms (redrawn from Williams and Hynes, 1977b).

pools. It is important to realize however that not all these methods will be open to the inhabitants of both temporary streams and ponds as, for example, in some streams the residual pools may dry up completely while in some regions burrowing crayfishes or equivalent tunnelers may not be present.

## 5.2 Variability in life cycles:

The drought survival method used reflects the type of life cycle exhibited by a particular species. Figure 5.2 gives examples of the

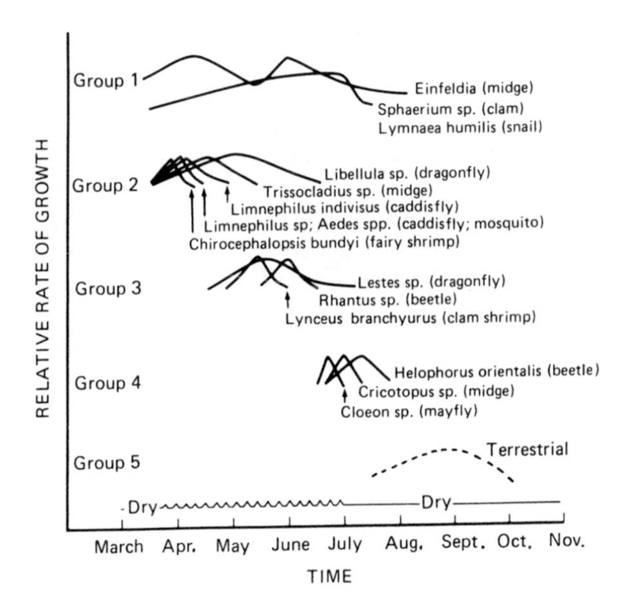


Figure 5.2: Features of the life cycles seen in the more common elements of the Sunfish Pond fauna in terms of the duration of generations and relative rates of growth (redrawn from Williams, D.D., 1983).

different types of life cycle shown by some of the more abundant animals in Sunfish Pond and illustrates relative growth rates. The species are grouped into five groups according to the seasonal succession outlined in Chapter 4 for the fauna of this pond.

In **Group 1**, the midge *Einfeldia* appears to have two periods of growth. Larvae in all four instars are present in mid April and the earlier ones grow rapidly until only third and fourth instars and pupae are evident in mid May. Adults are abundant by the end of May. The resulting eggs hatch quickly and the larvae grow rapidly in early June, passing into their second instar. Thereafter growth slows and the summer dry period is passed as early instars that can be recovered from the pond substrate. As soon as water reappears in the spring, the larvae continue their growth. Two of the common molluscs, *Sphaerium* sp. and *Lymnaea humilis* show gradual increases in growth throughout the pond phase with peak growth in the warm water of mid July just before dry-up. Virtually no growth occurs in the dry phase. Species in Group 1 are subject to the temperature range 0-27°C during the aquatic phase of their habitat.

The species assigned to **Group 2** show a variety of life cycles. Many, such as the fairy shrimp *Chirocephalopsis*, the caddisflies *Limnephilus indivisus* and *Limnephilus* sp. and the three species of *Aedes* mosquitoes grow very rapidly at the start of the pool phase. Although the life cycles of the *Aedes* species are very similar, all being univoltine and present as early instars concurrently, some staggering of development through late instars and emergence is evident. *A. trichurus* is the first to emerge, followed a week or so later by *A. fitchii* and finally *A. sticticus*. The midge *Trissocladius* shows a more gradual growth rate spanning a longer period - though it still emerges well before the end of the pool stage. Larval development of the dragonfly *Libellula* takes virtually the entire aquatic phase of the pond. Species in Group 2 are generally subject to less of a temperature range (0-17°C) during development.

**Group 3** species again show variable growth rates. The dragonfly *Lestes* grows gradually throughout much of the pond phase, while the clam shrimp *Lynceus ? brachyurus* and the beetle *Rhantus* grow more quickly, but at different times. These species grow in the temperature range 12-22°C.

Among the **Group 4** taxa, the mayfly *Cloeon* shows extremely rapid growth, completing its life cycle in 2-3 weeks. The midge *Cricotopus* grows quickly also, taking about 4 weeks, and completes its larval development in the moist pond substrate in mid July. The hydrophilid beetle *Helophorus orientalis* shows slower growth and passes much of its larval development in the moist bottom material. Water temperatures prevailing at the time of Group 4 development are in the range 17-27°C.

Suggested life cycles and growth patterns of the species of

**Group 5** are largely speculative as these primarily terrestrial animals were not sampled at other times of the year. It is likely that they grow at a reasonable pace given the warm temperature and plentiful food supply of the moist pond basin.

In temporary streams, there is the added dimension of moving water for part of the habitat cycle. This affects the life cycles of the fauna. Figure 5.3 shows examples of the life cycles of animals in Ontario temporary streams. There are three types of cycle evident amongst the fall-winter stream fauna:

**(1)** in which the eggs hatch immediately on the resumption of flow and the larvae develop quickly. Maximum size is reached just before ice cover brings cooler temperatures and the larvae remain at a low level of activity throughout the winter to emerge early in the spring, for example, the caddisfly *Ironoquia punctatissima*, the midges *Trissocladius* and *Micropsectra*, and the stonefly *Allocapnia vivipara*.

**(2)** in which the opposite occurs, that is, slow hatching and larval development at first followed by rapid growth as the water begins to warm up in the early spring. Most are mature before the stream stops flowing, for example, the microcrustaceans *Cyclops vernalis* and *Attheyella nordenskioldii*, the cranefly *Tipula cunctans*, the ostracod *Cypridopsis vidua*, the amphipod *Hyalella azteca* and the beetle *Agabus semivittatus*.

**(3)** in which there is a more even growth rate spread over the whole period of stream flow, for example, the flatworm *Fonticola velata*, the amphipods *Crangonyx minor* and *C. setodactylus*, the midges *Diplocladius*, *Orthocladius*, *Trissocladius* sp. and *Arctopelopia flavifrons* and the clam *Sphaerium*.

Similarly, three types of life cycle are evident amongst the spring-pool fauna:

**(1)** in which rapid growth occurs only at the beginning of the pool stage, for example, the mayfly *Paraleptophlebia ontario*, the beetle *Anacaena limbata* and the microcrustacean *Moina macrocopa*. This arrangement lessens the chances of these species getting "caught short" by an early dry-up.

**(2)** in which there is an even growth rate spread over the entire pool stage, for example, the waterstrider *Gerris*, the beetles *Helophorus orientalis* and *Hydroporus wickhami*, the dragonfly *Anax junius*, the backswimmer *Buenoa* and the fly *Tubifera*.

**(3)** in which multiple cycles occur in those species capable of very fast development, for example, the mosquitoes *Aedes vexans* and *Culex pipiens*. It is quite possible that in some cases, the last genera-

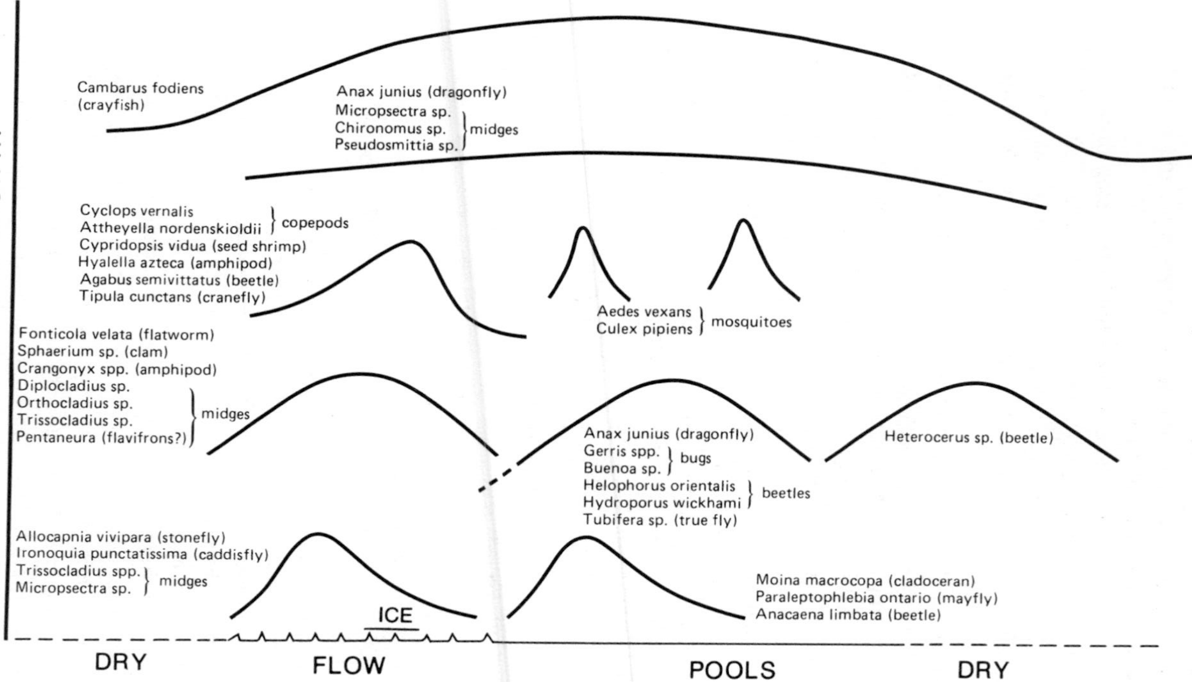


Figure 5.3: Diagrammatic representation of the different types of life cycle evident amongst some of the more common species found in Ontario temporary streams (redrawn from Williams and Hynes, 1977b).

tions of these species get wiped out each year as the pools finally dry out. However, mosquitoes have various special adaptations, such as staggered egg hatching, (a process that we shall examine in more detail later) which ensure the survival of at least part of the population. Growth rates for the summer-terrestrial fauna have not been studied, but it is likely that they follow the pattern of the mud-loving beetle *Heterocerus* which probably has an even growth rate throughout summer.

Two other life cycle patterns are obvious:

**(1)** such as shown by the flies *Pseudosmittia, Chironomus* and *Micropsectra*, and part of the *Anax junius* population, in which growth proceeds at a steady rate throughout the year with emergence probably taking place near the end of the pool stage.

**(2)** such as shown by the crayfish *Cambarus fodiens*, in which growth spans 2 years with peaks in development during the aquatic phases.

These variations in life cycle lead to the creation of a faunal succession from the stream stage through the pool stage to the terrestrial.

Stearns (1976) reviewed the ideas on the evolution of life cycles. He considered the key life history traits to be: brood size, size of young, the age distribution of reproductive effort, the interaction of reproductive effort with adult mortality, and the variation in these traits among an individual's progeny. He proposed several scenarios of traits that should evolve in species living in specified circumstances. Some of these scenarios address the specific problems of life history "tactics" in variable environments. He produced a classification scheme by considering the variability and predictability of the environment relative to the generation time of a population. Six classes became evident (Figure 5.4). *Class 1* - This is an environment in which regular, long periods of severe stress are interspersed with long periods when conditions are favourable for colonization, that is, long cycles where the species generation length is much shorter than the cycle period. Life history "tactics" here would be low age at reproduction, large clutch size, and parthenogenesis during the colonization phase, followed by either (a) sexual mating at the onset of stress to produce a diapause form capable of resisting the period of stress (e.g., cladocerans, other microcrustaceans and many aquatic insects) or (b) developmental plasticity in life history parameters leading to production of longer-lived iteroparous (multiple-reproducing) forms (e.g., some lymnaeid snails). *Class 2a* - This is

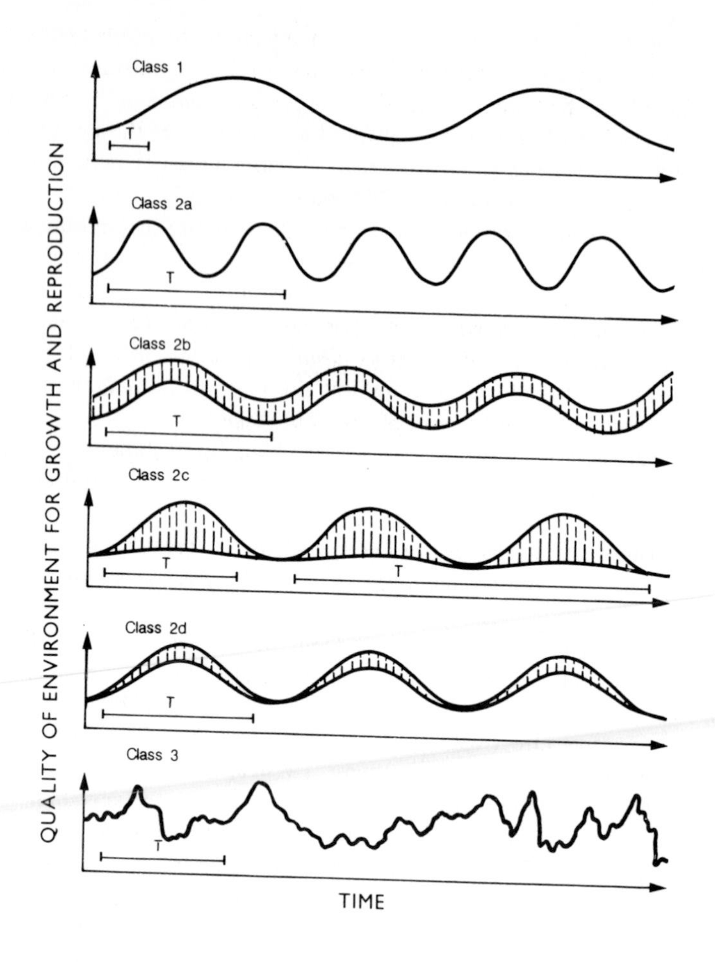


Figure 5.4: Classification of fluctuating environments (T = generation time and the lined areas represent regions of unpredictability in terms of either period or amplitude of fluctuation; redrawn from Stearns, 1976).

similar to Class 1 but the cycle period is shorter such that a life history spans one or more cycles. It is generally seen in large organisms living in stable, seasonal environments but also in organisms that either live or breed in the intertidal zone. As environmental conditions are predictable, the situation selects for breeding at the optimal time with little variance in breeding time (e.g., burrowing crayfishes and some fishes). *Class 2b* - Here the environment has cycles of intermediate length and the life history spans approximately one cycle. However, the starting time of the environmental cycle is unpredictable. The optimal "tactic" by an organism in this class of environment is to spread the risk of hatching too soon or too late in the cycle by developing a within-clutch polymorphism such that a distribution of hatching times in the clutch is generated which matches the historical probability distribution of the optimal date for hatching (eclosion)(e.g., mosquitoes). *Class 2c* - This is similar to Class 2b and the life history is usually the same length as or greater than an environmental cycle. The start of the cycle is predictable in time but the amplitude of fluctuations of environmental parameters is unknown. This situation characterizes species that live in habitats near the limits of their range where the possibility of a disastrous season in which no reproduction is possible becomes a reality (e.g., deserts and high latitudes). Populations may be maintained by recolonization from habitats with more predictable conditions. Variance in reproductive effort should be high and well correlated with variance in environmental quality. Such unpredictability in amplitude of environmental parameters select clutch sizes that are both smaller and more variable than the most productive size. Some species may evolve a "strategy" of retaining a certain proportion of offspring with delayed reproduction in order to minimize the risk of extermination. Species that make use of a dormant stage in the life cycle must strike a balance between the proportion of offspring emerging in any one year and the proportion dying during dormancy over the drought (e.g., beetles, hemipterans, midges, amphibians). *Class 2d* - Here the environment is similar to Class 2c and the length of the life cycle is equal to the length of the environmental cycle. The start of the environmental cycle is predictable but environmental fluctuations are partially unknown. One "tactic" in this type of environment is to postpone reproduction if conditions indicate a bad season ahead; this may involve resorption of the embryo and/or arrested development until better conditions prevail. It is seen mainly in large, long-lived vertebrates. Diapause "strategies" involve species locking in on environmental variables that are correlated with future conditions. Some so-called "annual" fishes can maintain permanent populations in temporary waters because the population survives the drought as diapausing eggs. Survival is based on the "multiplier" effect through which a single egg population of identical age can generate several sub-populations, each of which develops at different rates. This is similar to the phenomenon of staggered egg hatching seen in temporary water mosquitoes. *Class 3* - These highly variable environments have conditions that may be predictably favourable, predictably unfavourable or unpredictable during the habitable period. Stearns relegates most inhabitants of temporary ponds to this type of environment and predicts that when environmental conditions are favourable, the optimal "tactic" by a species in such a habitat should be rapid development and a total commitment of available energy to reproduction that produces a resting stage (e.g., most of the euphyllopods - the fairy shrimps, clam shrimps and tadpole shrimps). In the situation where conditions are unpredictable, a mixed "tactic" of producing some offspring that hatch at the first encounter with improving conditions and others at successive encounters, should be favoured by natural selection. Many temporary waters though, as we have seen, have a relatively predictable, cyclical occurrence so that they fit one of the other classes of variable environment (particularly Class 1) in Stearns' classification, and I think that the above examples bear this out.

## 5.3 Phenotypic and genotypic variation:

As the changes in water chemistry, temperature, dissolved oxygen, etc. of temporary waters are frequently great and rapid, the inhabitants must develop phenotypic plasticity of one genotype and/or maintain a considerable degree of genetic variability. A single species may be represented by several distinct ecotypes because of the uniqueness (chiefly in terms of physical/chemical properties) of each habitat (Bowen *et al.*, 1981). Adaptation to extreme environmental factors in individual waterbodies, together with inbreeding frequently leads to reproductive isolation (Templeton, 1980).

*Artemia salina*, the brine shrimp, is one of the most studied organisms from temporary waters particularly in terms of genetic variability and speciation which result primarily from adaptations to temperature and ionic composition (Collins and Stirling, 1980). *Artemia salina* is, as mentioned previously, a complex of many sibling species and semispecies and its taxonomic status is still in a state of flux. Populations are particularly adapted to chloride waters and the phenotypic plasticity which allows them to survive and breed in a wide range of anionic ratios represents the greatest known anionic tolerance of any metazoan species.

When populations of *A. salina* belonging to different sibling species are inter-mated there are no offspring. This complete reproductive isolation has been attributed either to a change in chromosome number or to parthenogenesis (Barigozzi, 1980). However, in a careful study of the *A. franciscana* cluster of non-parthenogenetic populations, Bowen *et al.* (1981) found incomplete reproductive isolation due to adaptations to habitat - each population being adapted to waters where the predominant anion might be chloride, carbonate or sulphate. They obtained viable $F_1$ and $F_2$ offspring from matings between these populations but pointed out that two populations which prove to be cross-fertile in artificial media in the laboratory may well be reproductively isolated in their native ponds because of the inability of one or both to complete their life cycles in the ionic composition or temperature regime of the alien habitat. These authors concluded, tentatively, that the phenotypic plasticity of a population of *A. franciscana* from Fallon Pond, Nevada was due to one genotype with a wide norm of reaction rather than to the presence of many genotypes.

It is clear that populations of temporary water species are particularly well-suited to studies of the processes of speciation and more emphasis should be placed on them in the future.

Sexual dimorphism is a phenotypic/genotypic trait common to many species. In the midge *Chironomus imicola*, flight is sexually dimorphic and has been related to selective pressures of its habitat (rain-pools that occur on isolated hillsides in tropical Africa) and to the disparate roles of flight between males and females. The intermittent nature and isolation of these habitats require a mandatory dispersal of gravid females during which they may have to fly for many kilometres. Consequently, females of this species show characteristics necessary for sustained flight, namely relatively long, broad wings that beat slowly and have a large amplitude of beat. Males, on the other hand have shorter, narrower wings that beat quickly with a short stroke, and their flight periods are typically shorter. These characteristics are particularly suited to aerobatic flight which is necessary for the successful acquisition of females in a competing swarm of males (McLachlan, 1986).

## 5.4 Physiological adaptations:

As we saw in Section 5.2, the suitability of ephemeral waterbodies as habitats for aquatic animals is highly varied, consequently many different types of life cycle "strategies" have evolved to enable

Table 5.1: Summary of physiological and behaviour mechanisms by which various aquatic organisms survive desiccation.

| Organism | Mechanisms/stage in life cycle | Reference |
|---|---|---|
| Algae/ Flagellate protozoans | modified vegetative cells with thickened walls; mucilaginous sheaths; accumulation of oil in cells; heat-resistant asexual cysts. | Evans, 1958; Belcher, 1970 |
| Sponges | reduction bodies, gemmules. | |
| Flatworms | dormant eggs; resistant cysts enclosing young, adults or fragments of animals; cocoons. | Castle, 1928; Kenk, 1944; Ball *et al.* 1981 |
| Rotifers | survive as dehydrated individuals; some bdelloids secrete protective cysts | Pennak, 1953 |
| Nematode worms | eggs; larvae; adults. | Pennak, 1953 |
| Clams | young and adult stages. | Thomas, 1963; Beetle, 1965 |
| Snails | adults form a protective epiphragm of dried mucous across shell opening; adults and young survive in moist air and soil under dried algal mats on pond bed. | Strandine, 1941; Eckblad, 1973 |
| Oligochaete worms | dormant eggs; resistant cysts enclosing young, adults or fragments of animals. | Kenk, 1949 |
| Leeches | survive as dehydrated individuals; some species construct small mucous-lined cells. | Hall, 1922; Pennak, 1953 |
| Tardigrades | resistant "tun" stage | Everitt, 1981 |
| Mites | ?resistant larvae; in most species, larvae attach to migrating insect hosts and leave the habitat, returning with the water. Larvae remain attached to host throughout its stay in permanent waterbody. | Wiggins *et al.*, 1980 |
| Fairy shrimp | resistant eggs. | Weaver, 1943; Hinton, 1954 |
| Tadpole shrimp | resistant eggs. | Fox, 1949 |
| Clam shrimp | resistant eggs. | Bishop, 1967 |
| Cladocerans (waterfleas) | ephippial eggs; adults survive in moist soil. | Elborn, 1966; Chirkova, 1973 |
| Copepods | diapausing eggs, late copepodites, adults (in cysts). | Cole, 1953; Yaron, 1964; Champeau,1971 |
| Ostracoda (seed shrimp) | resistant eggs; as near adults in moist substrate. | Pennak, 1953; Williams & Hynes,1976 |
| Amphipoda/ Isopoda | immatures near the groundwater table. | Clifford, 1966 Williams & Hynes, 1976 |
| Crayfishes | juveniles and adults in burrows at the groundwater table. | Crocker & Barr, 1968; Williams *et al.*, 1974 |
| Springtails | resistant eggs. | Davidson, 1932 |

*Continued*

Table 5.1: Summary of physiological and behaviour mechanisms by which various aquatic organisms survive desiccation.

| Organism | Mechanisms/stage in life cycle | Reference |
|---|---|---|
| Stoneflies | diapausing early instar. | Harper & Hynes, 1970 |
| Mayflies | resistant eggs. | Lehmkuhl, 1973 |
| Dragonflies | resistant nymphs; recolonizing adults. | Fischer, 1961; Daborn, 1971 |
| Hemiptera (true bugs) | recolonizing adults. | Macan, 1939 |
| Caddisflies | diapausing eggs; resistant gelatinous egg mass; terrestrial pupae in some species; recolonizing adults; larvae deep in substrate. | Wiggins, 1973; Williams & Williams, 1975; Imhoff & Harrison, 1981 |
| Beetles | semi-terrestrial pupae; burrowing adults; recolonizing adults; resistant eggs. | Jackson, 1956; Young, 1960; James, 1969 |
| Diptera (true flies) | | |
| Chironomid midges | resistant late instar larvae, sometimes in cocoons of silk and/or mucous. | Hinton, 1952; Danks, 1971; Jones, 1975; Stocker & Hynes, 1976; |
| Mosquitoes | resistant eggs; resistant late instar larvae and pupae of some species. | Schoof *et al.*, 1945; Horsfall, 1955 |
| Other dipterans | resistant eggs, larvae and pupae. | Hinton, 1953; Bishop, 1974; |
| Fishes | recolonization by adults; diapausing adults in substrate; resistant eggs. | Kushlan, 1973; Williams & Coad, 1979 |
| Amphibians | recolonization by adults; diapausing adults in substrate. | Wilbur & Collins,1973 |

species to exploit the resources of these environments. Much of the success of individual species in a particular type of temporary pond or stream is due to special adaptations of some aspect or aspects of the organism's physiology. The array of physiological adaptations that have arisen is almost as varied as that of the habitats themselves (Table 5.1). I propose to deal with examples under three broad groupings: behavioural avoidance, timing of growth and emergence, and diapause.

### 5.4.1 *Behavioural avoidance*

Several methods of avoiding the drought were indicated in Figure 5.1 and Table 5.1. Some species, notably adult insects and especially the beetles (Coleoptera) and true bugs (Hemiptera), fly away from drying waterbodies and spend the drought at nearby permanent waters, flying back upon refilling of the basin. Other adult insects may spend the dry period resting in sheltered places. In France, for example, the adults of the caddisfly *Stenophylax* that emerge from temporary streams fly to nearby caves where they hibernate under stable environmental conditions (Bouvet, 1977). Fishes also may migrate to and from temporary waters, especially tributaries that dry up, where the necessary connection exists with a permanent waterbody. We have already discussed the advantages that benefit suitably adapted fishes in ephemeral habitats.

Many species resort to burrowing to avoid the drought. For some (examples are given in Figure 5.1) this consists merely of crawling under debris (leaves, dry algal mats, etc.) or rocks on the bed. Others burrow to varying degrees into the substratum of the bed and most enter some form of resting or diapausing state. Many are invertebrate species but some vertebrates survive this way also, for example the tropical lungfishes and certain amphibians like the spadefoot toad.

Specialized freshwater crayfishes such as *Cambarus fodiens* can burrow to depths of a metre or more, the extent of the shaft depending on the depth of the groundwater table below the bed. As the groundwater table rises and falls so does the resting station of the crayfish (Figure 5.5). Pellets of mud from the excavation are brought to the surface and deposited as a chimney around the entrance to the burrow. In dry air conditions, these chimneys are plugged probably as a means of preventing moisture loss from the burrow. Most of the above activity takes place during darkness, especially on warm humid nights. Activity increases after heavy rainfall as this causes collapse of the chimneys and perhaps also the tunnels. During the fall, when the G.W.T. rises, the crayfishes start to move their resting stations towards the soil surface, and if the vertical shaft of the burrow is not too steep this merely entails climbing up as the water rises and establishing new or reoccupying old resting stations. However, if the shaft is perpendicular or almost so, the crayfish often fills in the lower parts of the shaft beneath it.

The water in the bottom of crayfish burrows serves as a refuge

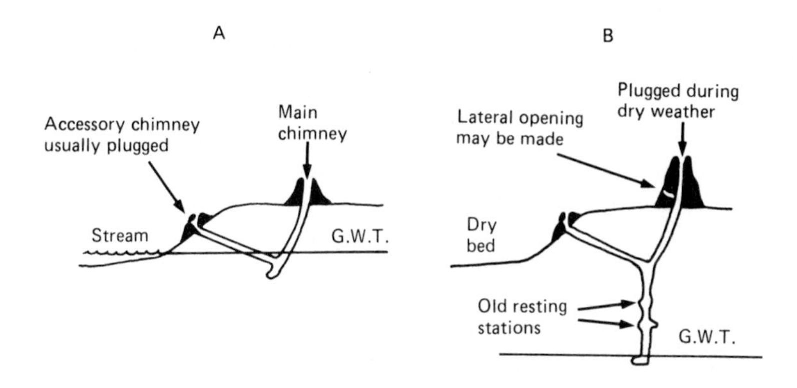


Figure 5.5: Diagrammatic sections through burrows of the burrowing crayfish *Cambarus fodiens.* (A) shows the typical Y-shaped burrow in early spring; (B) shows modification of the burrow during periods of receding groundwater table.

for smaller species and the term "pholeteros" has been coined to describe these assemblages (Lake, 1977). They have been recorded from burrows of crayfishes in North America and Australia and consist primarily of microcrustaceans, oligochaete worms, amphipods, isopods, springtails and midge larvae. The water in the burrow is generally of low pH and contains very little oxygen (e.g., 8-12% saturation), the pholeteros must therefore be physiologically adapted to these conditions. The crayfishes themselves are usually to be found at the water-air interface and it has been suggested (Grow and Merchant, 1980) that they may be extracting oxygen from the air in the burrow and not from the water (as long as their gills remain moist, crayfishes can obtain oxygen from the atmosphere).

Amphipod crustaceans, particularly the genus *Crangonyx*, and several genera of pond snails also have been observed to burrow in bottom sediments to avoid adverse conditions.

### *5.4.2 Timing of growth and emergence*

Figures 5.2 and 5.3 as well as showing the duration of species' life cycles in a temporary pond and stream indicate relative rates of growth throughout these cycles. The growth curves are of three basic types:

*(1)* a symmetrical (bell) curve indicates that during the animal's life cycle, the rate of growth first increases gradually and then decreases gradually as the animal approaches adulthood.

*(2)* an asymmetrical curve skewed to the left indicates a rapid rate of growth in the early instars, perhaps to take advantage of warmer water temperatures before the onset of winter or a temporary abundance of food, followed by a very slow growth rate thereafter. Such species will be in a near adult stage well in advance of any significant change in environment. This is a common type of growth curve in temporary waters. Some animals, such as mosquitoes, show such rapid growth that they are capable of completing two or more generations within one environmental cycle (Figure 5.3).

*(3)* an asymmetrical curve skewed to the right indicates a slow rate of growth for much of the life cycle, followed by a rapid burst of growth near the end - perhaps to take advantage of specific springtime environment conditions. Such cycles apparently are not flexible prior to significant change in the environment and this type of growth curve seems to be characteristic of a limited number of species.

Having now obtained a broad picture of the differences in growth rates between species, we should now consider some of the factors that control or appear to control growth.

*Water level* - Chodorowski (1969) found that dilution of pondwater by rain retarded the development rate of the mosquito *Aedes communis.* Both drying up of the pool and overcrowding in the population accelerated growth, the former effect being slight but the latter being great. McLachlan (1983) working with laboratory populations of the midge *Chironomus imicola* similarly suggested that patterns of adult emergence were associated with larval density, adaptively appropriate to the corresponding stage of evaporation of pools in nature. At initial low densities, safe, early emergence produced small females; increased risk for larvae emerging late was compensated for by achieving larger and therefore fitter adults. At high density, when the habitat was about to dry up, females emerged earlier but were also larger. Again, Fischer (1967) found that the development rate of the dragonfly *Lestes sponsa* accelerated in response to rapid water loss from a temporary pond; not that all populations of temporary water species have generation times perfectly synchronized with the end of the aquatic phase so that there is little or no wastage of reproductive effort. Hildrew (1985) found that in a temporary rainpool in Kenya the pool dried up on more than one occasion stranding many *Streptocephalus vitreus* (fairy shrimp) - many of which were still producing eggs. The strategy of this species seems to include extreme spreading of risk among the progeny, with rapid growth, early reproduction at a modest size and repeated mating to produce further clutches of eggs.

Broch (1965) studied the embryology of the eggs of the fairy shrimp *Chirocephalopsis bundyi* and the ecological and physiological factors controlling development and hatching. He found that embryological development was divisible into three blocks. Each was dependent on a set of ecological parameters associated with a woodland temporary pool and the development sequence was synchronized with the seasonal changes in these parameters. The first developmental block took place under high temperature and aerobic conditions during summer at the pond margins. This was termed the summer phase of synchronization and unless this stage of development was completed by autumn, there was no further development. The second block was a response to low temperature. This occurred in early autumn and under aerobic conditions resulting from the recession of water from the margin. The third block (prehatching to the metanauplius stage) took place in early winter at the dry pond margin. In the pond, prehatching began in a band around the periphery of the margin and this prehatching band moved to higher levels of the margin as winter progressed. Freezing conditions removed water, as a liquid, from around the eggs and inhibited prehatching. The unhatched metanauplii overwintered in a dormant state, until hatching was triggered by low-oxygen tension in the water at the soil-water interface when the pond filled in the spring. Thus it was the completion of embryonic development in a terrestrial environment from autumn to early winter that made synchronization of hatching with pond formation possible. Seasonal fluctuation of the water level in the pond was therefore the most important physical parameter in the early development of this and probably other species of Anostraca, although other parameters associated with the rise and fall of the water level were actually the controlling factors. In a study of the fairy shrimp *Streptocephalus vitreus* in the temporary rainpool in Kenya, Hildrew (1985) found that more eggs hatched from cores taken near the margins than at the centre. This ensures that the pool must be nearly full before the eggs are inundated and thus the chances of survival will be greater.

*Temperature* - Aquatic invertebrates are all poikilotherms, that is their body temperatures reflect those of their surroundings. As the rate of body metabolism is generally correlated with temperature, a rise in temperature increases the metabolic rate, invertebrates in temporary waters should grow faster during those phases when the water temperature is highest - provided that growth is not limited by some other factor such as available food or low oxygen levels. However, metabolism and growth rates will fall as upper (and lower) lethal temperatures are approached. If the relationship between metabolism and temperature were to be always as above, the life cycles and growth curves seen in any one temporary waterbody would be very similar for all species. Clearly, as we have seen, this is not the case. This may mean therefore that for temporary water faunas not all the species exhibit a simple relationship between temperature and rate of metabolism. Different species may have specific temperature ranges at which growth is maximal. Thus while certain temporary pond species grow faster in the low water temperatures after spring snowmelt, others grow faster in the high temperature water immediately prior to the summer dry phase. These differences contribute to the succession of species in the community discussed in Chapter 4.

*Food* - Growth of an organism is obviously affected by the quality and quantity of food available to it. We have seen (Chapter 4, section 4.5) that food webs in temporary waters include, primary producers, herbivores, detritivores and carnivores, in fact all the components of normal, efficient communities in other biotopes. Although seasonal availability of some of the food types contributes, along with temperature, to a temporal succession of some of the species in the community, many temporary waters have a plentiful supply of food particularly in the form of detritus. For those animals feeding on this detritus, Barlocher *et al.* (1978) have suggested that in ponds where decaying vegetation is exposed to air (in temporary vernal pools this would be during autumn and winter) the protein content, and thus the nutritional value of the detritus, in spring, will be greater than in permanent ponds. Enhancement of protein levels is the result of colonization of the detritus by fungi and bacteria which use the material both as a substrate for attachment and as a source of nutrients. The authors hypothesized that this protein-rich detritus in temporary pools might enable animals to grow faster, thus completing their life cycles safely before the pool dries up.

A very important aspect of the timing of growth and emergence is delayed and staggered egg hatching. Service (1977) studied the pattern of egg hatching in a population of the mosquito *Aedes cantans* from southern England. Figure 5.6 shows the weekly percentages that hatched from egg batches soaked continually for 32 weeks from October to May. Each year, eggs started to hatch in week 14

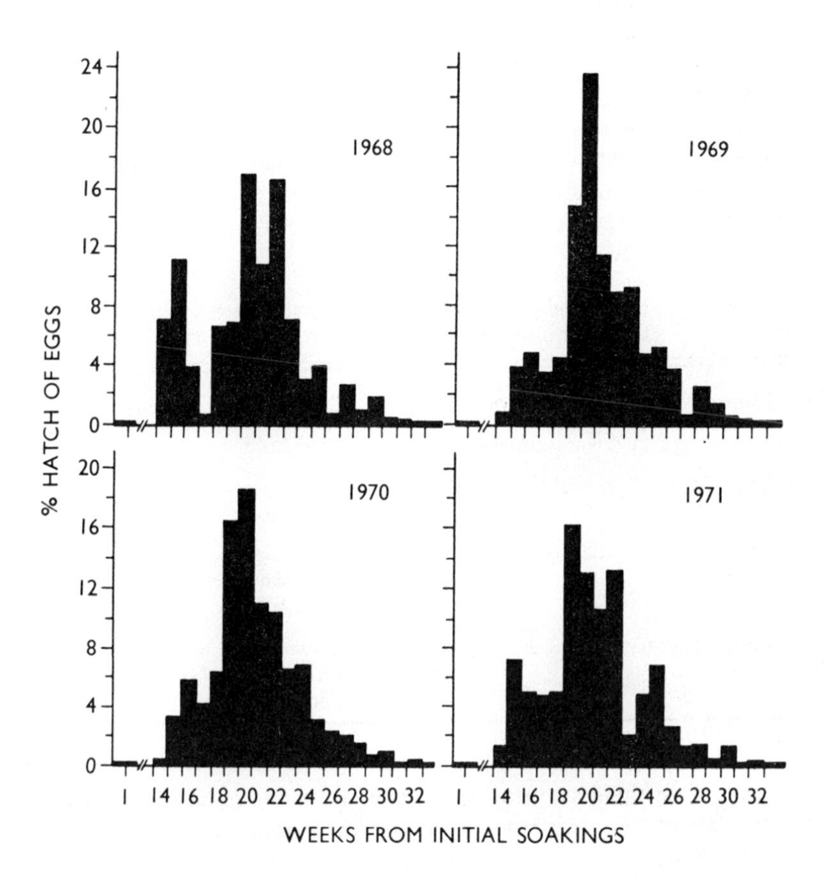


Figure 5.6: Weekly percentages of hatch of eggbatches of *Aedes cantans* first soaked at the beginning of October and then kept immersed for 32 weeks. Each year, eggs started to hatch in Week 14 (beginning of January)(redrawn from Service, 1977).

(January). By week 21, 50% had hatched and by week 28, 95%. The longest spread of hatching of a single batch of eggs was six days. Embryonic development required only 15 days after which, if relative humidity was maintained at 85% or higher, the eggs remained viable for many months - to a maximum of three and a half years if not flooded. Ponds in this locality were usually dry from May to September, however, if they became flooded during this period some eggs would hatch. From September to January no eggs hatched as they were in an obligatory diapause initiated by reduced temperature and/or day length. The factors terminating diapause were not determined, but Horsfall (1956), working on *Aedes vexans* found that once eggs of that species had been exposed to adverse environmental conditions they had to be conditioned before they would hatch. The conditioning process varied according to the eggs' prior environment (see Table 5.2). After conditioning, the basic hatching stimulus was a

Table 5.2: Examples of conditioning processes for diapausing eggs of the mosquito *Aedes vexans* (from Horsfall, 1956)

| Prior environment/Embryonic state | Conditioning sequence |
|---|---|
| Embryonated, in moist pond debris | Submergence, deoxygenation of water |
| Embryonated, in air-dried debris | submergence, de-watering, aeration of moist eggs, submergence, deoxygenation |
| Pre-embryonated, flooded | de-watering, aeration of moist eggs, submergence, deoxygenation |

decrease in the dissolved oxygen content of the pond water, caused for example by intense microbial growth. Perhaps by responding to such a stimulus, in nature, the emerging larvae may be preselecting temporary ponds that have already established an abundant food supply in the form of a healthy population of microorganisms.

Staggered egg hatching is not confined to the mosquitoes. *Limnadia stanleyana*, an Australian conchostracan (clam shrimp) lives in extremely ephemeral rainwater pools in eastern New South Wales. Bishop (1967) reported that generations of this species are frequently destroyed before maturing because the rainfall in this area is unpredictable in amount and irregular in occurrence; these habitats thus fall in the Class 3 environments proposed by Stearns (1976). As predicted, the eggs of this species are drought resistant but larvae are killed by drying. Thus, not all eggs hatch when the pools fill and many remain in the mud as a reserve to hatch at a later time. Darkness and low oxygen concentration in the mud inhibit hatching but each time the pools fill, some eggs are brought to the surface and hatch.

The fairy shrimp *Branchinecta mackini* occurs in temporary ponds in the Mohave Desert, California and populations may last from as little as three days to as long as four months depending on pond conditions. The populations survive drought as dehydrated eggs and hatch (within 24-36 hours) as water begins to fill the pond basin but only if the salinity of the water remains low (generally $< 1{,}000\mu$mhos). At low salinity, hatching is virtually continuous but, as salinity in these ponds generally increases rapidly after filling, the initial hatch is usually terminated quite quickly. Further episodes of hatching follow periods of influx of non-saline water as, for example, from rain or melting ice. For this species in these particular ponds, the duration of hatching is inversely proportional to the rate of increase in salinity until above $1{,}000\mu$mhos/day hatching is completely inhibited. In the laboratory it has been shown that egg hatching is actually controlled by a combination of salinity and dissolved oxygen operating in various proportions to either initiate or inhibit hatching. Geddes (1976) has similarly shown that the resting eggs of *Parartemia zietziana*, which occur in ephemeral saline ponds in Victoria, Australia, hatch in response to a salinity drop over a range of $50\text{-}200^{o}/oo$. From the results of their study, Brown and Carpelan (1971) proposed that the existence of a salinity-oxygen controlling mechanism might explain a difference between branchiopods of humid and arid regions. In both instances the animals are characteristic of temporary waters and these authors argued that stimulation of egg hatching must be due to some factor that changes at the time of origin of the pond. In humid regions, the factors seem to be temperature and oxygen. In arid regions, where the ponds may undergo significant changes in salinity, control of egg hatching may be by both salinity and oxygen, in which case, temperature simply controls the rate of development of the embryo.

### 5.4.3 Diapause

Table 5.1 gave some indication of the mechanics of drought survival across the entire spectrum of inhabitants of temporary waters. The degree to which each taxon survives the dry period is variable so some forms, for example the ostracods (seed shrimps) with their eggs enclosed by a double-walled shell of chitin impregnated with calcium carbonate, are more assured of success than others, for example hydropsychid caddisfly larvae that may reach a mortality of 50% of the population after only 15 days. Almost invariably the mechanisms that guarantee survival involve some form of diapause which may be defined as a period of suspended development or growth, which is accompanied by greatly reduced metabolism and is often correlated with season.

Diapause is sometimes closely associated with loss of water from body tissues. In its simplest form the relationship may be that lack of water arrests or retards growth such that the animal becomes dormant when dehydrated, but resumes growth whenever water is re-available. In many temporary water forms, therefore, the effect that hydration has on growth has been incorporated into the diapause mechanism (Lees, 1953). Along with a reduction in the water content of the body (below 50% for midge larvae) there is a reduction in the rate of oxygen uptake by the animal (Buck, 1965). Water loss during diapause can be dramatic in some species. A classic example is the midge *Polypedilum vanderplanki* which lives in temporary rock pools in Nigeria. These larvae can lose up to 99% of their tissue water but yet will recover on rehydration. In this state they can survive temperatures of 65°C for up to 20 hours, an adaptation that probably relies on an absence of water in the body (Hinton, 1951). Thoroughly dried eggs of *Artemia salina*, the brine shrimp, can be heated to 103°C for several hours and survive and whereas the upper lethal water temperature of the tadpole shrimp, *Triops*, is 34°C, the eggs can resist water temperatures close to 100°C (Carlisle, 1968). These extreme tolerances appear to be necessary, as, for example, the soil temperatures in dry pool basins in the Sudan can reach 80°C (Rzoska, 1984).

In North America, the nymphs of several winter stoneflies (especially species belonging to the families Capniidae and Taeniopterygidae) undergo diapause during the summer. This adaptation enables the species to inhabit streams that reach high temperatures or that dry up during the summer. The diapause usually takes place in an early instar and causes the nymphs to become morphologically distinct from other instars. Typically, the body of the inactive nymph becomes filled with fat globules which give it a characteristic pale colour; the head and antennae become reflexed under the thorax and the end segments of the cerci become thin and bald (Figure 5.7). The

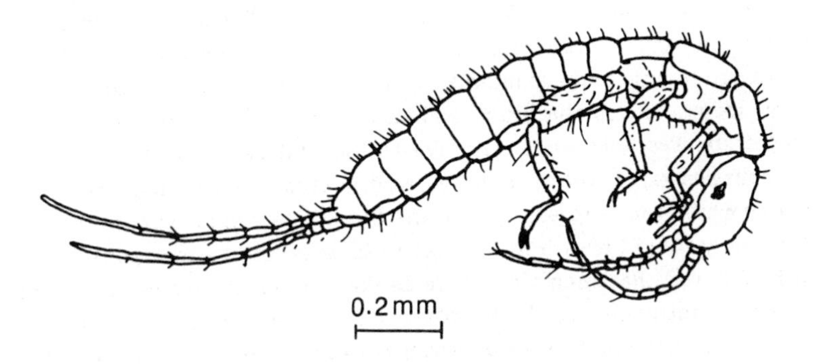


Figure 5.7: A diapausing nymph of the stonefly *Allocapnia vivipara* (redrawn from Harper and Hynes, 1970).

chief advantage of a nymphal diapause over an egg diapause is that it allows the nymph to seek out a suitable site for hibernation whereas a diapausing egg must lie where it falls. Frequently, the site chosen is deep within the substrate of the streambed, in the capillary fringe.

As we have seen, diapause as a means of surviving summer drought is by no means limited to insects. Elgmork (1980) summed up the adaptive significance of diapause in freshwater cyclopods as follows: it centres around two aspects. The first is an escape, in time, from unfavourable environmental conditions, thus it acts as a buffer in the life cycle allowing a species to inhabit niches that would otherwise not be available to it. The second concerns timing of important events in the seasonal cycle of the habitat, for example it may allow a species to take advantage of a unique, temporally-limited food supply before other non-diapausing species arrive. In addition diapause usually plays a role in synchronizing the reproductive phase of the species for the optimal time of year.

Even larger animals undergo diapause. One of the best-known examples is that of *Protopterus*, the lungfish which lives in temporary waters in Africa. As the water level in its habitat drops, *Protopterus* burrows into the bottom mud and secretes a cocoon of mucus around itself. This cocoon is made waterproof by a layer of lipoprotein and, inside, the fish lies folded upon itself with its head adjacent to a small opening at the top of the cocoon. Inside the cocoon of *P. aethiopicus* the fish's rate of oxygen uptake is gradually reduced to approximately 10% that of an active fish. Heartbeat drops to about three beats/minute and the fish loses some water from its tissues. Part of the cocoon extends into the fish's pharynx and this acts as a respiratory tube. Urine production stops and nitrogen excretion changes from producing toxic ammonia to urea which, being less toxic, can then accumulate in the blood and tissues with no adverse effects. In this state, *Protopterus* can survive for several years. Termination of diapause is rapid, within minutes, and is stimulated by immersion in water which tends to asphyxiate the fish which cannot survive in water solely by gill-breathing - it has to have access to atmospheric air. It is believed that such a rapid "re-awakening" must involve release of a special hormone into the bloodstream (Beadle, 1981).

Another lungfish, *Lepidosiren paradoxa*, which lives in the temporary swamplands of central South America, survives drought in a similar manner.

In Australia, several species of gudgeons (Gobiomoridae) and native minnows (Galaxidae) are known to bury themselves in mud or moist soil as their habitats dry up, but little is known of their physiology or adaptations.

In the copepod crustaceans, diapause is not restricted to habitats that dry up, in fact it was first recorded for species that inhabit large, deep lakes. Here, animals migrate into the profundal depths where they may burrow into the sediment. When diapause ends, the juvenile animals migrate back up into the warmer, more productive upper layers of the lake where further development and reproduction occur. It is quite possible that this form of diapause was a preadaptation that favoured the establishment of copepods in temporary waters as diapausing stages are usually much more tolerant of extreme variation in a number of environmental parameters.

The stage in the life cycle in which diapause takes place is variable. In the cyclopoid copepods it is most frequently as copepodite IV (there are five such stages before the adult stage, and six nauplius stages precede the first copepodite stage). In calanoid copepods, diapausing eggs have been found, whereas in harpacticoid copepods diapause is most common in adults or late copepodites. The harpacticoid *Attheyella nordenskioldii* (Figure 5.8) secretes a cyst around itself and diapauses near the groundwater table. The cyst itself consists of a single layer of a transparent, homogeneous gelatinous substance to which sand grains become attached. It requires several

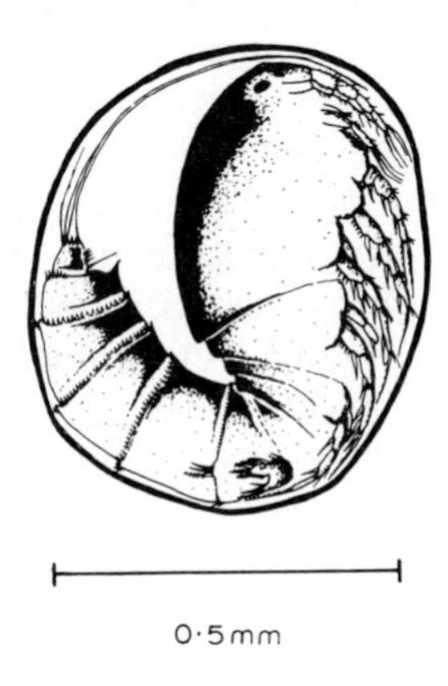


Figure 5.8: Summer drought-resistant cyst of the harpacticoid copepod *Attheyella nordenskioldii* (redrawn from Williams and Hynes, 1976).

hours of immersion in water before the animal can break out and swim away. In some species it appears that high water temperature, rather than desiccation, is the stimulus that initiates diapause. As temporary waterbodies usually warm up before drying out, cyst formation as a means of surviving desiccation could well be controlled solely by temperature rise, and this could be a legacy from the possible origins of copepod diapause in large, permanent waterbodies. Factors that break diapause in copepods have not been well studied but they do not appear to be as complex as, for example, those breaking the egg diapause of mosquitoes. In the oligochaete worm *Tubifex tubifex*, cyst formation can be induced simply via desiccation (the function of the glandular clitellum shifting from egg cocoon formation to cyst formation). Equally as easily, excystment can be induced by immersion in water for 20 hours. Kaster and Bushnell (1981) have proposed that the ability of this species to form short-term protective cysts may be a prime factor in its success in disturbed areas as well as accounting for its cosmopolitan geographic distribution. The same may be true for other, similarly adapted species.

We have seen that high temperatures may induce diapause, what of low temperatures? Many waterbodies at high latitudes and altitudes form temporary habitats for aquatic species by virtue of the fact that the water they contain freezes solid for part of the year. Danks (1971) has shown that chironomid midges in the Arctic construct firm, generally elongate cocoons, sealed at both ends, that are composed almost entirely of salivary secretion but sometimes include mud particles. In this state the larvae are capable of surviving, frozen solid in the mud, as these ponds freeze to the bottom for many months each year. The larvae become noticeably shrivelled due to loss of some water from their body tissues and, just as we saw for species surviving high temperatures in tropical regions, this partial dehydration seems to enhance resistance to temperature extremes - in this case freezing. Fully-hydrated larvae fail to complete their development due primarily to destruction of their fat body (Leader, 1962). Asahina (1966) pointed out that insects susceptible to freezing are killed when the temperature of their environment corresponds to their supercooling points. In the case of terrestrial insects, it has been shown that the presence of glycerol in the animal's tissues causes increased survival through depression of the supercooling point. The presence of this polyhydric alcohol, and also sorbitol and mannitol, has been discovered in the diapausing, overwintering larvae or eggs of several species (Somme, 1964). Olsson (1981) has described overwintering of several taxa in a north Swedish river that freezes solid in winter. Many of the animals survived enclosed in some form of covering. For example, the worm *Lumbriculus variegatus* produced transparent spherical cysts in which the animal lay tightly folded; many chironomids constructed the elongate cocoons already described; the snail *Gyraulus acronicus* formed an epiphragm over its shell opening; and many of the caddisfly larvae blocked both openings of their cases even though they were not in pupal or pre-pupal stages. It is possible that these various devices served as mechanical protection against the physical stress of ice crystals forming in the surrounding environment.

Whenever diapause occurs, it is typified by a profound metabolic arrest which allows survival through adverse environmental conditions on the organism's endogenous reserves. Characteristically, the organism ceases its consumption of external nutrients concomitant with a metabolic depression of variable but controlled degree (Hochachka and Somero, 1984). Alongside this occur programmed sets of biochemical adjustments for protecting the organism against low or freezing temperatures and/or against lack of food and water. The period of diapause may be relatively short as, for example, in species occurring in temporary autumnal pools, or it may be extensive, as in species occurring in temporary vernal pools, where it frequently occupies a significant fraction of the organism's life cycle.

## 5.5 Physiology of desiccation:

Many temporary water species exhibit not diapause in the strict sense, but anhydrobiosis which, although similar, differs because in the latter, metabolism is brought to a complete standstill. Anhydrobiosis involves the ability of the organism to survive loss of all cellular water (perhaps with the exception of that bound tightly to macromolecules) without sustaining irreversible damage (Hochachka and Somero, 1984).

Anhydrobiosis is quite a widespread phenomenon and is seen in most major taxa. Crowe (1971) has subdivided these into two groups: **(1)** organisms capable of anhydrobiosis only in their early developmental stages - including spores of bacteria and fungi, plant seeds, eggs and early embryos of some crustaceans (e.g. *Artemia*, the brine shrimp), and the larvae of certain insects; **(2)** organisms capable of anhydrobiosis during any stage in their life cycles - including certain protozoans, rotifers, nematodes and tardigrades. These groups share many similar biochemical mechanisms concerning initiation and termination of anhydrobiosis. Hochachka and Somero (1984) gave two examples of anhydrobiotic systems - that of *Artemia* and that of nematodes, which I shall summarize here:

*Artemia* - the *Artemia* cyst consists of a non-cellular shell surrounding an embryo in the early gastrula stage. This embryo consists of approximately 4,000 undifferentiated cells which have no obvious morphological features that indicate them to be desiccation-adapted. Despite this, the embryo can tolerate virtual complete loss of water yet remain viable for many years. Table 5.3 shows the relationship between the water content of *Artemia* cysts and metabolic activity, as outlined by Clegg (1981). Below 0.3g of water present/gram of initially dehydrated cyst material (g/g) no metabolic reactions take place, and any chemical transformations that do occur may do so without the contributions of enzyme catalysis - this is the ametabolic domain. Some metabolic activity may begin at hydration levels near 0.3 g/g but this is not reflected by a measurable increase in oxygen uptake and consists of only a few types of metabolic reactions - this is the domain of localized or restricted metabolism and spans the hydration range of 0.3 to 0.65g/g. Above 0.65g/g, oxygen consumption rates become detectable and rise with increasing water content. Metabolic activities here are qualitatively the same as those found in fully-hydrated cysts - this is the domain of conventional metabolism. Restoration of cellular water results in rapid reactivation of metabolic and developmental processes.

Table 5.3: Hydration-dependence of cellular metabolism in the cysts of the brine shrimp *Artemia* (after Hochachka & Somero, 1984).

| Cyst hydration ($gH_2O$/g cysts) | Metabolic events initiated | |
|---|---|---|
| 0 to 0.1 | None observed | ametabolic domain |
| 0.1 | Decrease in ATP concentration | |
| 0.1 to 0.3 ± 0.05 | No additional events observed | |
| 0.3 ± 0.05 | Metabolism involving several amino acids, Krebs cycle and related intermediates, short-chain aliphatic acids, pyrimidine nucleotides, slight decrease in glycogen concentration | domain of restricted metabolism |
| 0.3 to 0.6 ± 0.07 | No additional events observed | |
| 0.6 ± 0.07 | Cellular respiration, carbohydrate synthesis, mobilization of trehalose, net increase in ATP, major changes in the free amino acid pool, hydrolysis of yolk protein, RNA and protein synthesis, resumption of embryonic development | domain of conventional metabolism |
| 0.6 to 1.4 | No additional metabolic events observed | |

Clegg proposed that the initial addition of water to the cysts is needed to form initial hydration layers around macromolecules. Some of this "bound" water may have remained in the cyst even under extreme desiccation, but hydration levels of 0.15g/g allow full repletion of this store. Further hydration (up to 0.6g/g) may replenish the "vicinal" water component of the cyst - this is water which is relatively "organized" within the cells and is associated with longer range interactions with membrane surfaces and macromolecular complexes such as cytoskeletal elements. Yet more hydration (0.6g/g +) is thought to contribute to the "bulk" water, that is, the water associated with other, general, cellular components. The bulk aqueous phase appears to be necessary to allow effective transport of metabolites, fuel sources, and the like, between cellular compartments.

There is a strong correlation between survival in the desiccated state and accumulation of polyols (polyhydroxy alcohols) in the cells. Polyols seem to be important in osmoregulation and resistance to freezing and may be necessary in anhydrobiosis as well. Two polyols are evident in cysts of *Artemia*: glycerol comprises about 4% of cyst dry weight, while trehalose accounts for up to 14%. In anhydrobiosis, polyols may act as water substitutes, creating hydrogen-bonded interactions with polar or charged entities of the cell, and, in addition, they may stabilize protein structure at low water activities. The latter occurs because glycerol is excluded from the highly structured water surrounding proteins. The addition of glycerol therefore promotes the compact, folded structure of proteins thus preventing them from unfolding and becoming denatured. When water returns to the cell and there is a need for rapid reactivation of metabolism, the necessary enzymes will be at hand in a functional state (Gekko and Timasheff, 1981).

*Nematodes* - in roundworms, anhydrobiosis is not confined to any particular stage in the life cycle. As in *Artemia* cysts, the stimulus for anhydrobiosis appears to be reduced availability of water. This begins a series of progressive morphological changes which take place between 24 and 72 hours after initiation. The entire animal contracts longitudinally and becomes coiled, intracellular organelles such as muscle filaments undergo ordered packing, and membrane systems show ordered change. The sequence of events is initiated by water loss, but the morphological changes themselves are under the endogenous control of the animal itself. The main metabolic adjustment that occurs during dehydration is one of redistribution of carbon from storage products such as glycogen and lipid into large intracellular pools of trehalose and glycerol. In nematodes these two polyols reach levels of 10 and 6%, respectively, of the animals' total dry weight. Similar changes in polyhydroxy alcohols are known to occur in other anhydrobionts. Glycerol may be important in stabilizing membranes in Group 2 organisms (Crowe, 1971).

There are few data available on the metabolic events at different stages of hydration in the nematodes but it can be tentatively assumed that, as in the cysts of *Artemia*, metabolism comes to a halt during anhydrobiosis.

During rehydration of desiccated nematodes, metabolic events appear to be the reverse of those leading to entry into anhydrobiosis although the rate at which they happen is very different: up to 72 hours for dehydration but only a few hours for rehydration. This is presumably linked to the relative times the pond takes to dry up (long) and refill (short), and thus is significant in terms of the effective timing of occurrence of the species in the pond.

Anhydrobiosis seems to bestow many advantages upon those species that have managed to incorporate it into the biology. These advantages are: *(1)* enhancement of dispersal, particularly by wind, *(2)* allowing, or improving chances of, survival in habitats subject to severe drought, *(3)* synchronization of biological processes (e.g. feeding, growth, reproduction, etc.) with favourable episodes in the environment, *(4)* increased longevity, *(5)* preadaptation for survival of other adverse environmental factors (e.g. temperature extremes and anoxia).

# 6 COLONIZATION PATTERNS

## 6.1 Introduction:

This chapter deals primarily with those animals that avoid the drought phase of a temporary pond or stream by leaving the habitat. Later, when the water returns, they either recolonize the same habitat or colonize a similar one within their dispersal range. Many such species use temporary waters exclusively as feeding and breeding grounds and spend the rest of their time in permanent water bodies.

## 6.2 Adaptive "strategies" of colonizing animals:

What factors make good colonizers? According to Lewontin (1964), the species should be capable of effective dispersal, have high somatic plasticity (i.e. respond readily, in body form at least, to changes in environmental conditions) and have high interspecific competitive ability. Few species possess all these attributes to a high degree and often it is a case of finding habitats where one may be exploited. For example, many species which are poor interspecific competitors have high powers of dispersal. In this way they are able to colonize newly-flooded habitats first - either annually or seasonally - produce a quick, new generation and then emigrate before the arrival of species that may be competitively superior but which colonize at a slower rate or before the habitat dries up again. However, in temporary waters the brevity of the aquatic phase may prevent much competition from occurring and success therefore depends more on power of dispersal and plasticity of the life cycle. Once the former has enabled the species to reach the habitat, the latter may help to maintain it there.

Species that devote much of their energy to dispersal are often referred to as fugitive or pioneer species and animals that show this trait can be likened to the "weed" species of plants. MacArthur and Wilson (1967) put forward the idea that natural selection takes different directions depending on whether a species is living under crowded or uncrowded conditions. Species living in uncrowded conditions are termed *r*-selected species while those living near the carrying capacity (that is near the asymptote of the logistic growth curve - see Figure 6.1) of the population are termed K-selected species. *r*-

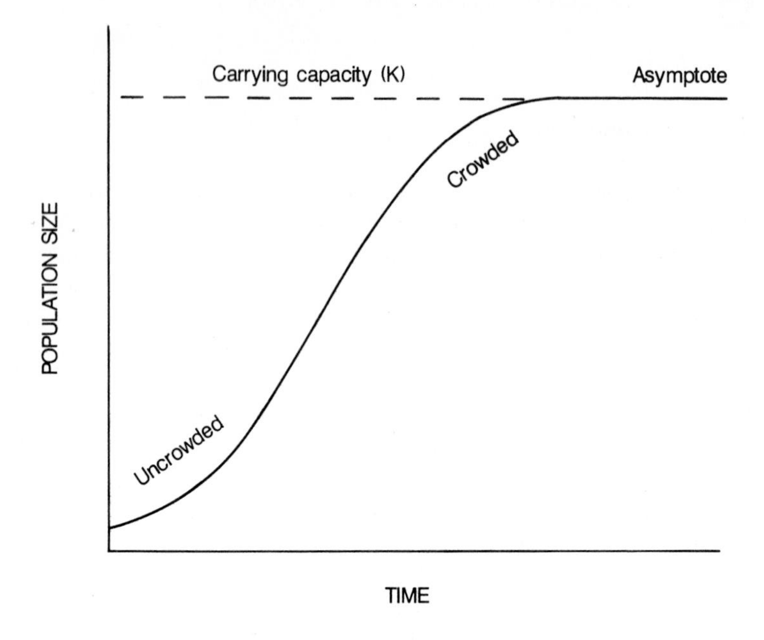


Figure 6.1: Logistic growth curve for a species.

selected species are well suited to living in temporary or unstable habitats while K-selected species fare better in stable habitats; a comparison of the main attributes of these two species types is given in Table 6.1. Clearly then, the majority of species colonizing temporary waters should be *r*-selected. However, recent work by Istock *et al.* (1976) on the mosquito *Wyeomyia smithii* which lives in the water-filled leaves of the pitcher plant - a temporary water microcosm - has shown that larval populations alternately experience selection that is density-dependent and density-independent according to available resources. This species' position on the *r*-K continuum thus varies with time. The authors concluded that colonization of a highly specialized, monophagous resource may result in adaptive strategies following this resource more closely than other species in larger temporary waters where a variety of food resources are available and where *r*-selection and rapid growth are the norm.

Life history adaptation appears to become more complex in

Table 6.1: Comparison of features of *r*—and K—selected species

| *r* – selected | K – selected |
|---|---|
| – typically found in unstable habitats | – typically found in stable habitats |
| – high powers of dispersal | – low powers of dispersal |
| – poor competitive ability | – good competitive ability |
| – high intrinsic rate of natural increase | – lower intrinsic rate of natural increase |
| – large number of eggs | – small number of eggs |
| – early reproduction in life cycle | – later reproduction in life cycle |
| – small body size | – large body size |
| – short life span (e.g. annual) | – longer life span (several years) |
| – opportunistic/generalistic feeders | – specialized feeders |

temporary waters that are highly saline. Williams (1985) produced a habitat template in which the two axes are the extent of predictability of the habitat and salinity levels (Figure 6.2). The figure shows the presumptive distribution of three broad types of selection recognized by ecologists: K (or interactive selection) and *r* (or exploitative selection) we have seen already, but also A, or adversity selection. The figure again shows the importance of *r*-selection in temporary waters. However, no organism is entirely *r*-, K- or A- selected as, generally, a balance is struck which maximizes the adaptive value of features drawn from each type of selection (Pianka, 1970).

All animals have the capacity for dispersal and it may take one of two forms, active or passive. The former involves the animal reaching a destination "under its own steam", so to speak - usually by flight or crawling, while the latter relies upon the animal being transported by some external agent such as the wind or a larger animal. Although a greater element of chance in reaching a suitable new habitat occurs in passive dispersal/colonization, as we shall see, it is not entirely out of the control of the transportee.

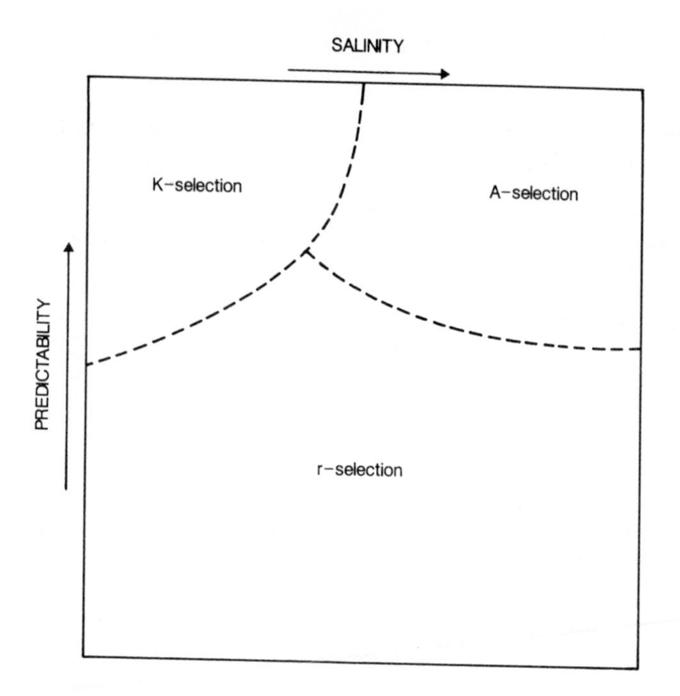


Figure 6.2: Habitat template showing the distribution of *r*–, K– and A– selection according to habitat salinity and predictability (redrawn from Williams, 1985).

## 6.3 Active colonization:

This is largely restricted to those insects that possess good powers of flight, particularly the beetles (Coleoptera) and true bugs (Hemiptera), but also the dragonflies (Odonata) and the true flies (Diptera). However, non-insects such as crayfishes, leeches, amphipods and amphibians have been known to crawl overland between nearby ponds.

Some of the earliest observations on insect colonization were made in Britain. Grensted (1939), for example, erected a small (1.5 m diameter) open-air canvas tank in his garden "with no higher scientific purpose than the amusement of small children in hot weather". Within 24 hours of having been filled with tap water, it had been colonized by over one hundred water beetles representing 8 different species although *Helophorus brevipalpis* was the most common.

Since his garden was a "long way" from water he concluded that these species flew freely in suitable weather and then mostly in the afternoon and evening. This occurrence was confirmed by the Reverend E.J. Pearce (1939) who was "shewn a small birdbath, in Sussex, that had been repaired and filled only a day or two previously and which could not hold more than a couple of gallons when full. It was literally swarming with *Helophorus brevipalpis*. There must have been several hundred specimens present". The nearest permanent water was a lake about one-third of a mile away.

### *6.3.1 Flight periodicity:*

Aerial colonization of temporary waters is limited to the spring and summer months in temperate latitudes, although it may occur all year round in the tropics if not controlled by the monsoons. The best studied groups have been the Hemiptera and Coleoptera. Macan (1939) found that six species of Corixidae (waterboatmen) were commonly on the wing in Cambridgeshire and that colonization of water bodies was frequent. He showed that a succession of species occurred depending on the percentage of organic matter present in the ponds and he hypothesized that high mobility was necessary to maximize niche separation. Knowlton (1951) noted a flight of corixids from one pond at 8:40 p.m. and their landing at another pond at 10:00 p.m. Temporary pond species migrated more frequently than those living in permanent waters, and immature specimens seemed more inclined to fly during dispersal times.

Fernando (1958) working at Oxford showed that emigration of aquatic beetles from one pond and their subsequent colonization of others occurred seasonally. He found that *Helophorus brevipalpis* had two distinct dispersals in Britain, one in the spring (April-May) and the other from the middle of June to the end of August, the size of the second being much greater than that of the first (Figure 6.3). Those females that took part in the spring dispersal carried mature eggs which were laid in the newly colonized habitats, while the late summer flight was a dispersal one. Working later in Canada, Fernando found two similar dispersals for *Helophorus orientalis* and another beetle *Anacaena limbata* but their maximum seasonal abundances never coincided - when *Helophorus* ceased its colonization flight in early July, *Anacaena* began its summer dispersal. With the decline in flight of *Anacaena* at the end of August, there was another flight by *Helophorus*. These staggered flights were probably the result of

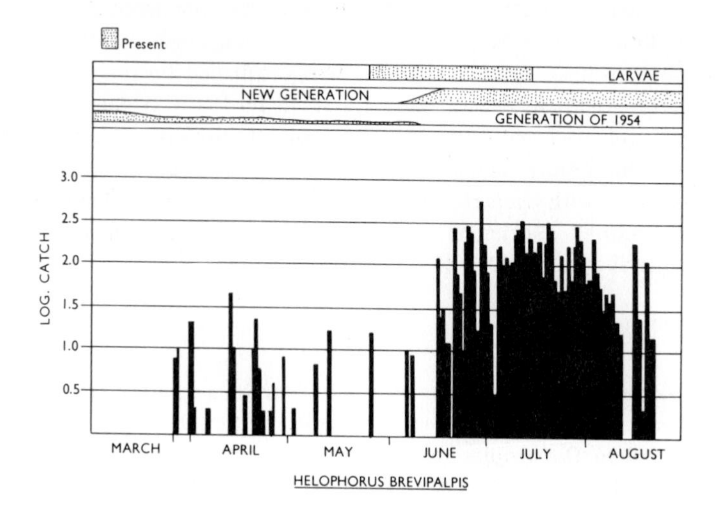


Figure 6.3: Logarithm of the catches of *Helophorus brevipalpis* in the field experiment in Wytham during 1955, using six artificial habitats (4 ft x 4 ft), three glass traps (5 ft x 3 ft), and three glass traps (2 ft x 3 ft) — a daily census was made from March to September (redrawn from Fernando, 1958).

different responses to climatic factors and may have resulted in reduced competition between the two species for the same habitat. In addition, a bimodal, daily flight pattern with approximately equal morning and evening peaks was evident for these two beetle species. Landin (1968) observed a similar pattern for *H. brevipalpis* in Sweden but, as Figure 6.4 shows, the number of beetles captured during the evening was 3 to 4 times that captured during the morning. Temperature may determine when an insect species flies through fixed values constituting thresholds for this activity. *Helophorus brevipalpis* has a low temperature threshold of between 11 and 15°C and there is some evidence for an upper threshold of around 25°C (for the Canadian species the range is 15-27°C). Beament (1961) has shown that the permeability of the cuticles of the water beetles *Dytiscus* (diving beetle) and *Gyrinus* (whirlygig beetle) to water increases 6-fold above a cuticle surface temperature of 23°C. If similar values apply to *Helophorus* and *Anacaena* (and a great many aquatic insects have cuticles

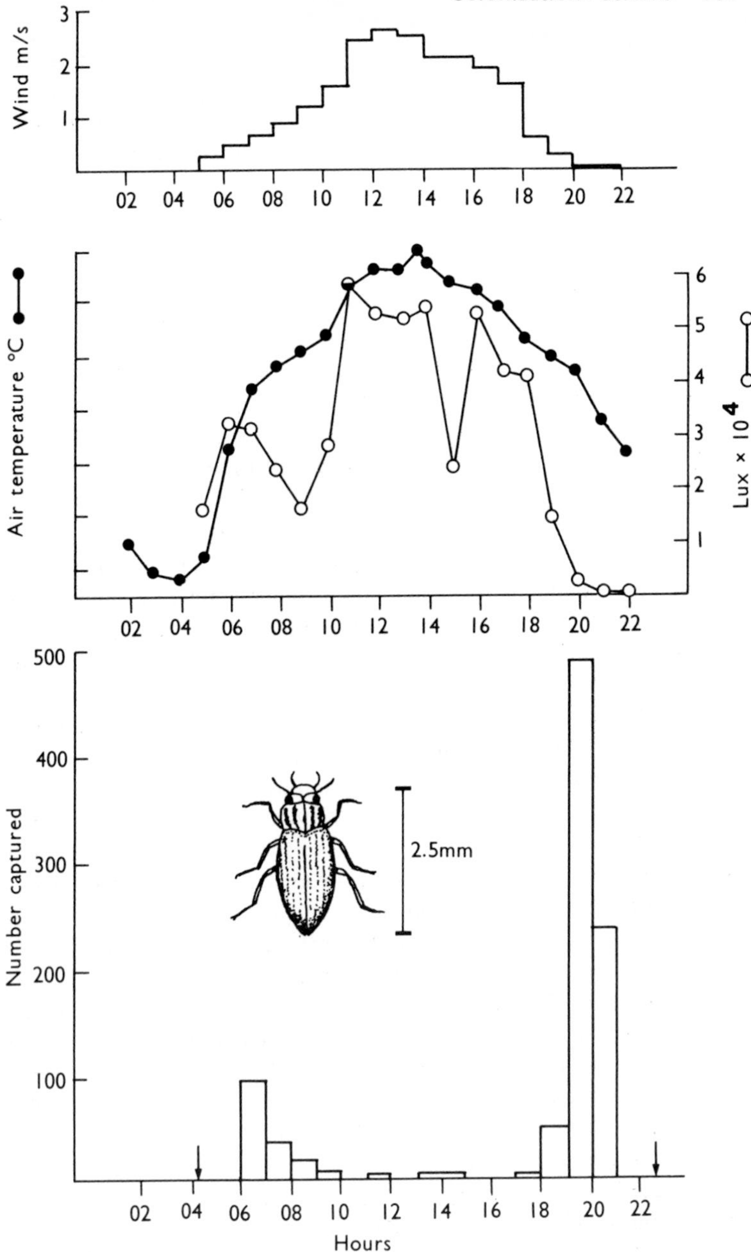


Figure 6.4: Flight periodicity in *Helophorus brevipalpis,* together with air temperature, light intensity and wind velocity on the 22 July, 1965 (wind velocity measured from 5 am to 10 pm sunrise 3.12 am, sunset 8.34 pm; redrawn from Landin, 1968).

which change permeability rapidly at low temperatures) then it is likely that the reason for diurnal periodicity in flight is that summer midday temperatures exceed that which allows normal cuticle permeability thus subjecting the beetle to dramatic loss of water from its tissues.

### *6.3.2 Flight initiation and termination:*

Many factors control the pattern of insect colonization of temporary waters. Fernando (1958) believed colonization in aquatic beetles to be the end product of a series of steps in each individual's behaviour. These steps are: *(1)* dispersal, which provides the basis for colonization, *(2)* location of habitat, and *(3)* selection of a habitat. The steps are products of evolution caused by two types of factors, proximate factors like temperature and light, and ultimate factors like food. Beetles respond directly to the proximate factors by adaptations of physiology and behaviour, thus water or air temperature change, change in light intensity and duration, rain or a combination of these factors may cause them to leave the water and take to the air from where they may locate a suitable new water body. It is possible also that internal physiological states, such as ovarian development in the female, or hunger, initiate flight, as Landin (1980) has found that, in the spring, flying *Helophorus brevipalpis* females have larger oocytes than non-flying ones and the guts of fliers are invariably empty. Ultimate factors such as food and substrate may not be directly connected with the proximate factors but the insect may often anticipate their being favourable for itself or even its offspring because of the evolution of seasonal dispersal.

Flight terminates with pond selection and either colonization by the adult insect itself or egg laying by the female to establish a new generation. The actual process of pond selection may involve one or more of the insect's senses. Popham (1953), for example, showed that sight was important in corixids, as flying adults orient themselves at a set angle to the incident light and thus the reflection of light from a water surface is sufficient for them to "home-in" on. This characteristic explains why many species of aquatic bugs, beetles and dragonflies swoop down and land on the shimmering hot surfaces of tarmacadamed roads or the surfaces of freshly polished automobiles. Pajunen and Jansson (1969) observed that adult corixids could discriminate pool size, moving from shallow pools in which they lived in the spring, to pools deep enough for over-wintering in the late autumn. Water movement, shade, colour, alkalinity, salinity and emergent vegetation all have been implicated in the habitat selection/oviposition stimulation processes of temporary water mosquitoes.

Most remarkable is the process by which many temporary water insect species and amphibians select and oviposit in ponds and streams when they are dry. How does the female "know" that the dry bed upon which she lays her eggs will later develop into an aquatic habitat suitable for sustaining aquatic larvae? It seems that colonizing females may be attracted by some special feature of the bed. For example, a moisture gradient emanating from the soil water or the scent of rotting aquatic vegetation have been suggested as attracting gravid mosquitoes. Corbet (1963) thought that female dragonflies might be attracted by a specific preference for a plant such as *Typha* which only grows where the ground is susceptible to seasonal flooding.

### 6.4 Passive colonization:

Unlike species that are capable of controlled, active colonization, those that rely on passive means are subject to a much greater degree of chance in populating new habitats. Although adult and larval stages are sometimes involved, it is more common that a resting, reproductive body acts as the transportee. A large proportion of freshwater species have such a stage in their life cycles, which is not the case for marine species. Talling (1951) cites examples of these stages: algal spores, sponge gemmules, polyzoan statoblasts and the resting eggs of rotifers, fairy shrimp, waterfleas, copepods and flatworms. This same author also outlined the factors governing the chance element in colonizing ponds (Table 6.2).

Agents of passive dispersal can be assigned to two broad groups: the first is abiotic in nature, the second, biotic.

Very light organisms can be transported by air currents, for example the parachute seeds of *Typha*, and strong winds acting on the drying beds of temporary waters can sweep away viable algal cells and the ephippia of waterfleas. Rzoska (1984) proposed that the hot winds of the Sahara Desert sweeping over and picking up the dry debris in shallow pond basins, are responsible for spreading the resistant stages of many taxa, particularly those of the Notostraca, Conchostraca, Anostraca, Cladocera and Copepoda. Maguire (1963) examined the colonization of small, water-filled jars by these aerial forms and concluded: that the number of different organisms per bottle decreased with increased height above the ground, that they similarly decreased with increased distance from the source pond, and that levelling off of the numbers of different organisms was approached after six weeks near the ground but after only two weeks in

Table 6.2: Factors governing the chance element in the colonization of ponds and streams (from Talling, 1951)

A. Factors governing the dispersal of reproductive bodies:
1. Intrinsic properties of the organism
   (a) formation of drought-resistant reproductive bodies, e.g. cysts, ephippia, eggs, reduction-bodies.
   (b) capacity for active overland dispersal (minimal in passive forms)
   (c) frequency of occurrence in nature (density and patchiness)
2. External agencies of dispersal
   (a) wind
   (b) inflowing stream and water drainage
   (c) animals – particularly waterfowl and adult insects
3. Local factors
   (a) age of pond or stream
   (b) area of pond or stream
   (c) distance from similar habitats

B. Factors governing successful colonization by the reproductive bodies:
1. Intrinsic powers of multiplication and competitive ability
2. "Openness" of the habitat (accessibility in terms of degree of cover)
3. Destruction by other organisms (e.g. predation)

the bottles kept four feet above the ground. Disseminules frequently were washed into the bottles from nearby vegetation and soil surfaces by rain. A practical investigation of passive aerial colonizers, based on Maguire's experiment, is given in Chapter 8.

Temporary waters that are periodically connected to streams may be colonized by invertebrates such as midge larvae, snails and aquatic worms which are rafted in on algal mats that have broken off from the bed.

Charles Darwin was amongst the first to observe passive colonization through biotic agents in the form of organisms attached to ducks and waterbeetles. Since then there have been numerous records of various taxa found attached to flying waterfowl. For example, Proctor (1964) recovered viable eggs of fairy shrimps, tadpole shrimps, clam shrimps, waterfleas and ostracods together with algae from the lower digestive tracts of wild ducks; and amphipods have been seen clinging to the feathers of dead mallards that were some distance from, and had been out of water for several hours. An interesting correlation has been made between the distribution of sibling species of the brine shrimp *Artemia* and bird migration routes (Bowen *et al.*, 1981).

Smaller hosts may contribute also to the passive dispersal/colonization of certain groups: algae, protozoans and aquatic fungi have been collected from flying adult bugs, caddisflies, craneflies, midges, mosquitoes, aquatic beetles and dragonflies. Fryer (1974) found bivalve molluscs attached to the leg hairs of waterboatmen (corixid bugs), and ostracods (seed shrimp) often attach to notonectids (backswimming bugs). Viable eggs of the brine shrimp, *Artemia salina*, have been recovered from the faeces of crayfish species with which they coexist, and I have seen crayfishes with small clams clamped tightly on the tips of their walking legs walk several metres before shaking them off.

One of the best examples of a habitual passive colonizer, however is the water mites. The larvae are parasitic on insects and have

Figure 6.5: Waterboatman (Corixidae) with a heavy infestation of hitchhiking watermite larvae.

good organs of attachment in the form of specially modified claws and mouthparts. Favoured hosts are the Hemiptera (true bugs), especially waterstriders and waterboatmen, but mosquitoes and midges are parasitized too, frequently as pupae (Figure 6.5). As the pupa rises to the water surface to emerge into the adult insect, the mites very quickly transfer their hold from the sloughed pupal skin to that of the imago and are thus carried away from the pond by the insect. By selecting host species specific to temporary waters, the mites are virtually assured of successful transference to a suitable new habitat.

## 6.5 Colonization and competition:

In some temporary aquatic habitats, for example rain-filled rock pools in tropical Africa, the fauna is qualitatively poor and, in fact, may consist of just one or two species. This begs the questions of how this comes about and, perhaps more interestingly, how these virtual monocultures are maintained? (McLachlan and Cantrell, 1980). Recall that three species of dipteran larvae predominate in these pools: *Chironomus imicola* in larger pools that last for several weeks before drying; *Polypedilum vanderplanki* in smaller pools that are very short-lived; and *Dasyhelea thompsoni* in pools of intermediate size and duration.

In the case of the two chironomid species, pools either contain *P. vanderplanki* or *C. imicola*, rarely both, and they do not replace each other in a seasonal succession within the same habitat. McLachlan (1985) produced evidence to show that "possession" of a particular pool is not determined by the first species to arrive, nor is it the result of non-equilibrium conditions. The outcome is determined by interacting variables, some of which are physical in nature, the others biological. McLachlan ranked the physical variables of pool location, duration of the water-phase, and the amount of bottom sediment as primary factors. Biotic variables, including competition (based on body size), population density, accumulation of "metabolites" and stirring up of the sediment by tadpoles, were ranked as secondary but still important. Because experimental introductions of *P. vanderplanki* larvae into pools normally inhabited by *C. imicola* failed, McLachlan ruled out chance (see earlier ideas of Talling, 1951; Table 6.2) as a causal factor. *Polypedilum vanderplanki* seemed to be excluded by scarcity of sediment (in which they normally shelter), the presence of tadpoles (which disturb the sediment), and the presence of "metabolites" (= essence of) of *C. imicola*, which condition the water in some way that discourages colonization or establishment of *P. vanderplanki. Polypedilum vanderplanki* persists only in pools where there is enough sediment, where there are no tadpoles, and where the pool basin is shallow enough to be flushed by rain, thus removing the "essence" of *C. imicola.*

The two species only really compete in short-lived pools. Here, because *P. vanderplanki* survives dry periods as partially-grown, diapausing larvae, as soon as the pools are refilled it reappears and, being larger than the first instar larvae of *C. imicola* (which cannot tolerate drought and have to recolonize via eggs laid by migrating females), it can out-compete the latter. *Chironomus imicola* shows many of the attributes of an extreme *r* strategist, while *P. vanderplanki,* perhaps because of its unusual tolerance to desiccation, is able to establish itself over the longer term in these rainpools and may therefore show a few of the attributes of K-selected species.

## 6.6 Temporary waters as islands in time and space:

The equilibrium theory of island biogeography (MacArthur and Wilson, 1963; 1967) states that isolated oceanic islands are continuously being colonized by new species while some previously-arrived species are lost or become extinct from the biota. The fauna and flora *per se* may never be static, in terms of species complement, but, at some time, the immigration rate will equal the extinction rate and a state of dynamic equilibrium will be attained. This theory has been applied not only to oceanic islands but to other habitats isolated in space such as coral heads, baskets of sterile rocks on the beds of streams, and the faunas of mountain peaks. It also appears to be applicable to "islands" that are isolated in time such as the aquatic stages of temporary waters.

Ebert and Balko (1982) produced a standard graphic model of colonization and extinction rates, showing the equilibrium points for species' numbers, modified for time rather than space (Figure 6.6). They equated frequency of occurrence of a habitat in time with distance from a source in space arguing that patches of habitat, such as isolated ponds, are a long distance in time from a source of possible colonizing species and that resting stages (disseminules - spores, cysts, eggs, seeds, etc.) may have to travel for considerable periods of time between favourable habitat conditions. As in the original model for islands in space, "islands" distant in time have lower rates of colonization than do "islands" that are closer (i.e. temporary waters

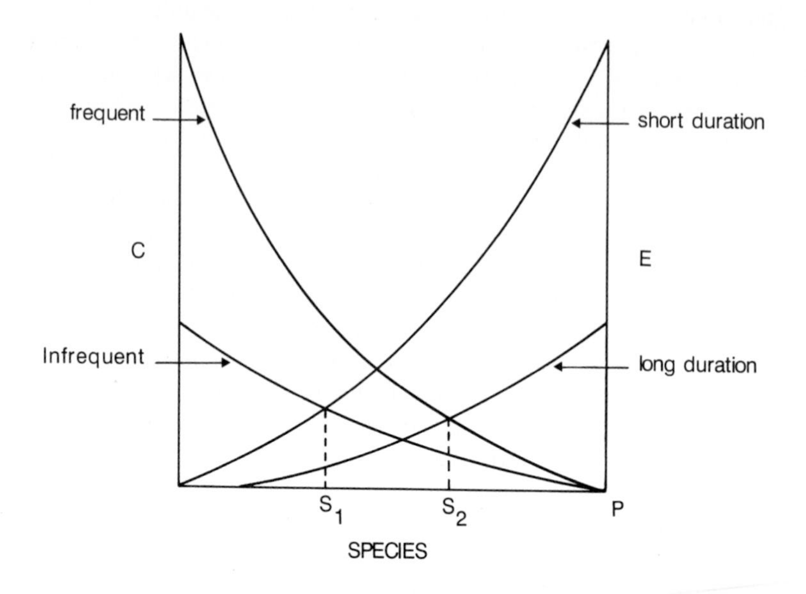


Figure 6.6: Diagram of colonization (C) and extinction (E) curves for "islands" in time indicating the analogues with "islands" in space. P = the number of species in the species pool; S1 represents the number of species at equilibrium for an "island" that occurs infrequently and has a short duration; S2 represents the equilibrium number of species on an "island" habitat which occurs frequently and has a long life (redrawn from Ebert and Balko, 1982).

that fill more frequently). In the original space model, rates of extinction were shown to be related to the physical size of the island whereas, in the time model, the analogue is duration of the habitat. Vernal pools that dry up frequently may be considered to be islands of short duration, having a small size in time, while permanent water bodies are islands of long duration and are large with respect to time.

The time model (Figure 6.6) predicts that infrequent, short-lived water bodies will have fewer species at equilibrium ($S_1$) than frequently-occurring water bodies of longer duration ($S_2$). However, many species living in temporary waters are affected by both time and space factors and thus a more complete model should accommodate both. Ebert and Balko therefore proposed the following, tentative

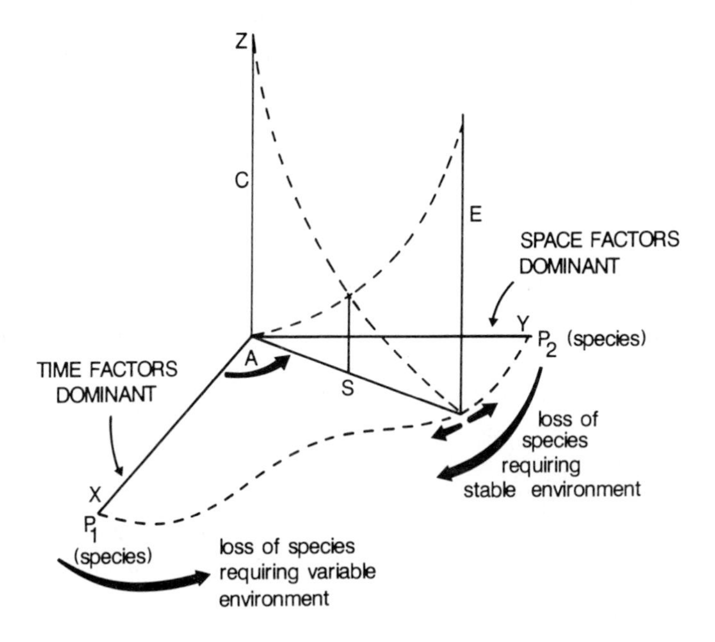


Figure 6.7: Three-dimensional representation of colonization (C) and extinction (E) rates (the Z-axis) on a species plane. Points on the line from P1 to P2 represent the realized species pools that can supply colonists to "islands". A realized pool near the X-axis would be for a species group that perceives habitats more like "islands" in time than in space; a realized species pool nearer the Y-axis would be for species groups that perceive habitats more as "islands" in space (redrawn from Ebert and Balko, 1982).

model (Figure 6.7). The graph shows both components of time and space with axes for colonization and extinction rates drawn perpendicularly to the X-Y species plane. Two extreme temporary waterbodies are shown on the X- and Y-axes as $P_1$ and $P_2$. The X-Y plane represents the species pool of all possible colonists, within an individual taxonomic group, onto all possible types of "islands", and there would be different X-Y planes for each taxonomic group being considered. Because of temporal instability the waterbody may be suitable for some species but not for others. Some species will be part of species pools over much of $P_1$ to $P_2$, indicating that they have very broad tolerances/adaptations and normally occur in many habitats from those that are highly ephemeral to those that are permanent. Others have very narrow tolerances and occur over a very restricted region of the species plane.

Rotation from $P_1$ to $P_2$ on the graph represents movement along a gradient of temporal habitat instability. The most ephemeral habitats are nearest the X-axis. As one approaches the more permanent waters, nearest the Y-axis, species that are highly adapted to temporary water tend to drop from the species pool but species that can only survive in permanent waters are added to the potential species pool. Rotation from $P_1$ to $P_2$ not only describes the change in potentially colonizing species but represents a gradient of the influences of temporal versus spatial factors of the habitat. On the X-axis, the rates of colonization and extinction are largely determined by time factors such as the frequency and duration of the habitat. On the Y-axis, colonization and extinction rates are largely determined by spatial factors such as the size of the waterbody and the distance from a source of potential colonizers. The rates of colonization and extinction between $P_1$ and $P_2$ are functions of frequency, duration, size and distance.

As in the original concept of the equilibrium theory as proposed by MacArthur and Wilson, the equilibrium number of species (S) is, on Ebert and Balko's "species plane", the point where the rates of colonization (C) and extinction (E) are equal. The angle, A, is a function of the duration and frequency of the waterbody such that:

$$A = j(f,u)$$

"j" will have a maximum value of $\pi/2$ when "f" (the frequency of habitat occurrence) and "u" (habitat duration) are both equal to 1.0. That is, the species pool is $P_2$ in a permanent waterbody - which, by definition, occurs each year and survives for the entire year, with a probability of 1.0. By including A in the determination of S, it is possible to account for changes in species composition as well as changes in species number.

There are a few cautionary notes to be added to the application of this innovative, multidimensional model, however. There may well be other factors besides space and time, that influence the rates of colonization and extinction, for example weather (Donald, 1983), food quality and availability, and interactions between species. Further, as the pool of colonists may vary seasonally this may prevent some habitats from ever attaining a true equilibrium (see Williams and Hynes, 1977a; Williams, D.D., 1980).

# 7 OTHER TEMPORARY WATER HABITATS

## 7.1 Introduction:

The purpose of this chapter is to briefly consider those temporâry aquatic habitats that have been somewhat neglected by researchers even though they are commonplace. A few, such as treehole habitats and the chambers of insectivorous pitcher plants have recently received more attention and thus will be dealt with in more detail.

Leewenhoek (1701) studied the faunas of roof gutters and concluded that the animals he found, predominantly rotifers and tardigrades, could be completely dried out without losing vitality - an obvious advantage to living in such a precarious environment.

The animals living on the shores of permanent lakes and ponds in which water levels fluctuate widely are also subject to many of the same adversities as temporary water inhabitants. There are many examples of wide fluctuation of water level in natural lakes, particularly in the tropics. Lake Chilwa in Malawi, Africa, is a prime example. It lies in a tectonic depression and covers an area of approximately 700km$^2$. At high water, it spreads outwards to form flooded marshes that are dominated by *Typha domingensis* (bulrush) and *Aeschonomyne pfundii* (ambatch). In recent times the change in level is of the order of 1 to 3m, vertical height, but the positions of ancient beaches indicate that, in prehistoric times, fluctuations may have been as great as 30m. The peripheral water is less saline than that of the main basin and supports a benthic fauna consisting primarily of chironomid larvae, dragonfly and mayfly nymphs, leeches and pulmonate snails. Of the 30 fish species recorded from the lake, only three are common in the open lake. Many fishes take refuge in swamps and scattered pools around the basin in times of drought (Beadle, 1981). The importance of swamps and floodplains will be examined in more detail in Chapter 8.

Hynes (1961) studied the effects of such fluctuations on the littoral fauna of Llyn Tegid in Wales. This is a lake used for flood control and the study was made soon after this was implemented, a time when the water level range was about 5m. The study showed that many of the original littoral species were wiped out but that new and dominant species arrived that could tolerate these conditions. Upon later reduction in the degree of water level fluctuation, some of the temporary water forms were replaced by some of the original permanent water species (Hunt and Jones, 1972). As waterbodies become more and more subject to water level control by man, increased occurrence of this form of temporary aquatic habitat is inevitable. Ward and Stanford (1979) make reference to more than 12,000 dams greater than 15m high that have been constructed on the world's major rivers. The total surface area of reservoirs thus created is in excess of 300,000km$^2$. In these situations, temporary water habitats will be created both on the shores of the reservoirs and on the margins of the receiving rivers.

On a much smaller scale, there is a large number of temporary waterbodies collectively known as "container" habitats. These include natural forms such as the leaf axils of plants (especially of tropical bromeliads); teasels; bamboo habitats - formed when chrysomelid beetle larvae bore into large bamboo stems; the decaying pulp of split cocoa pods; rat-gnawed coconuts; split paw-paw stems; empty snail shells; and cup fungi (Mattingly, 1969). Some of these habitats are extremely small. The total volume of water contained in the bract of a pineapple plant, for example, may be as little as 10cm$^3$, yet it may support populations of two species of mosquito (Barton and Smith, 1984). These water-retaining structures formed by terrestrial plants are called *phytotelmata* and are found on at least 1,500 species of plants belonging to 29 families. Although phytotelm habitats contain species representing many of the major aquatic insect orders, the true flies (Diptera) are the most common with more than 20 families reported. It is estimated that some 15 genera, containing 400 species, of mosquito inhabit these waterbodies, and although they show varying degrees of host-plant specificity, few species also occur in non-phytotelm habitats - this may be due to chemical oviposition cues of plant origin (Fish, 1983).

In coconut habitats, there is a succession of mosquito species as the organic content of the water decreases and is diluted by rainwater. Mattingly (1969) stated that these rat-gnawed coconuts are important breeding sites for mosquitoes of the *Aedes scutellaris* complex, which are vectors of the debilitating filarial worm which causes human elephantiasis. Such is the considered importance of these habitats to the mosquitoes that control of rats is a recognized ancilliary procedure for control of filariasis in the Pacific.

Man-made microhabitats also harbour important tropical vectors. For example, the mosquito *Aedes aegypti* breeds in water that collects in rainbarrels, cisterns, tin cans, jars, iron cooking pots and old motorcar tyres; it transmits yellow fever.

Many of the species that are to be found breeding in container habitats specifically seek out these small habitats. Buxton and Hopkins (1927) illustrated this in their study of the factors influencing oviposition of *Aedes polynesiensis* and *A. aegypti* on Pacific islands. Apart from finding that the females' egg laying was influenced by such factors as light and darkness, wind, rain, landing places, temperature, water vapour and salinity, they also found a preference for small containers. More eggs were collected from containers 15cm in diameter than from a series of smaller and larger containers. It is the egg stage, in particular, that exhibits moderate drought resistance in container-inhabiting species (Lounibos, 1980).

Again, these micro-aquatic habitats are not the exclusive domain of insects. Muller (1880) first made the discovery of microcrustaceans in the axils of Brazilian bromeliads and since then many, particularly ostracods, have been recorded from a variety of container habitats (Victor and Fernando, 1978).

For insects, at least, container habitats may be ancient environments. Barton and Smith (1984) summarized the evidence for this along the following lines: *(A)* much of the fauna is derived from the more primitive lines within aquatic insect taxa, or the fauna is of tropical origin. This is especially true of the Diptera - most of the families that are successful in container habitats belong to the suborder Nematocera and many temperate zone species are derivatives of large, tropical groups. The pitcher plant mosquito of the nearctic (*Wyeomyia smithii*), for example, belongs to the tribe Sabethini, which is a successful and widely distributed group in the tropics. *(B)* Breeding in container habitats is cosmopolitan in occurrence, having evolved independently in many taxa and in many regions. *(C)* Ecological studies to date (although limited in number) indicate, particularly for the Diptera, a number of morphologically and behaviourally similar trends which suggest a long association of species and habitat. *(D)* High specificity for certain types of container habitats by some species, e.g. *Wyeomyia smithii* are found only in pitcher plants.

## 7.2 Treeholes:

A treehole may be defined as any cavity or depression existing in or on a tree. Kitching (1971) designated two types, "pans" which have an unbroken bark lining, and "rot-holes" which penetrate through to the wood tissue of the tree (Figure 7.1). Both types collect

Figure 7.1: Example of a treehole habitat at the base of a maple tree.

leaf litter but rot-holes contain, in addition, a layer of fungus-rotted wood. They occur in a wide variety of trees, especially deciduous ones. Few aquatic animal groups are represented in treehole communities and these include only a limited number of insect orders. The most common elements of the fauna are dipterans and beetles. Nematocerans such as mosquitoes, chironomids, biting midges, craneflies, moth flies and wood gnats are the most frequently encountered dipterans, while marsh beetles (Helodidae), scirtids and pselaphids (mould beetles) are the most common beetles. Occasionally, other taxa are found, for example, mites. Fashing (1975) found *Naiadacarus arboricola* exclusively in treeholes of a number of tree species in Kansas. Small crustaceans, rotifers and protozoans have been recorded also (Rohnert, 1950).

Rohnert (1950) proposed a classification scheme for the faunas of treeholes based primarily on their degree of specificity towards these habitats: **Dendrolimnetoxene** - this includes small taxa such as protozoans and rotifers that are readily dispersed, passively. Their occurrence in treeholes may be largely accidental. **Dendrolimnetophile** - this includes aquatic and semi-aquatic species that may be found in other habitats such as forest floor litter, pools and marshes. Examples are various families of Diptera, woodlice, and air-breathing snails. **Dendrolimnetobionten** - this includes primarily insect species, such as belong to the Diptera and Coleoptera, that are obligate inhabitants of treeholes. This classification can also be applied to other container habitats.

Kitching (1971) found that in Wytham Woods, England, the inhabitants of treeholes fed saprophagously and had a variety of life cycles. Some, like the beetle *Prionocyphon* took two or more years to complete a generation, while the midge *Metriocnemus* completed up to three generations each year. Most species spent the winter as quiescent larvae or adults and the densities of all species were higher in holes in the forest canopy layer than closer to the ground; the latter was attributed to better quality detritus in higher holes.

The probable mechanism for partitioning the detritus among the fauna of the Wytham Woods' treeholes is illustrated in Figure 7.2. The two species feeding on large particle detritus have different habits - the helodid beetle breaks down leaf particles within the detritus layer, while the syrphid fly grazes on particles on the surface of the detritus. Potentially, the two midges seem to compete for small particle detritus but the larvae are temporally segregated - the chironomid is most common in the summer, while the ceratopogonid is most common in the winter. Competition between the two mosquito species is avoided because *Aedes geniculatus* can exploit small particle detritus as well as organic matter in suspension, and it therefore has a food refuge. In addition, *Anopheles plumbeus* is comparatively rare and, in practice, the two species seldom coincide (Kitching, 1983).

Small habitats that are temporally and spatially variable, and thus unpredictable, should select for species that are phenotypically

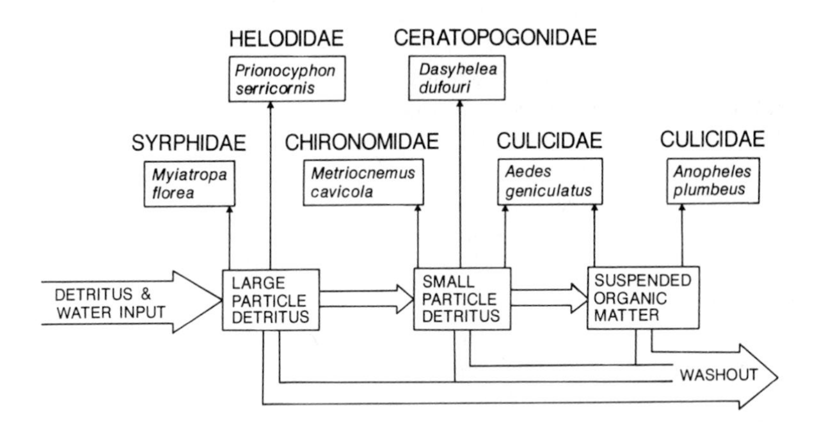


Figure 7.2: Foodweb for the treehole community of Wytham Woods, England (redrawn from Kitching, 1983).

plastic, genetically polymorphic, and that exhibit frequent reproduction (iteroparity) and dispersed oviposition (Giesel, 1976). Watts and Smith (1978) showed much of this to be true in a study of *Toxorhynchites rutulis*, a predatory, treehole mosquito. This species shows precocious oogenesis in that females emerging from the pupal skin have almost mature eggs. This is a feature common to many short-lived insects but *T. rutulis* has a long life span, compared with other mosquitoes. Ovarian precocity must insure that some eggs will be deposited in the treehole in which the female was raised. Coupled with this, *T. rutulis* exhibits oviposition behaviour that can be interrupted and which enables the female to lay between 1 and 7 eggs in a number of different treeholes. In addition, reproductive effort by females is continuous and asynchronous, rather than the periodic and synchronous pattern found in most other nematoceran Diptera. Presumably, this allows the female of *T. rutulis* to have a reserve of eggs that may be deposited whenever these small aquatic habitats become available.

Some interesting information on the similarity of treehole habitats and their faunas between different parts of the world has been provided by Kitching (1983). He compared the microclimate of treeholes in Wytham Woods with those in Lamington National Park, Queensland, Australia. In Wytham Woods, a mixed, deciduous forest, the treeholes were situated most commonly among buttress roots and branches. The average surface area of each reservoir was 423cm$^2$ and they were, on average 25.2cm deep. pH was around 6.4 and the mean conductivity was 339$\mu$S/cm. The treeholes received an annual leaf litter input equivalent to approximately 934g/m$^2$. Lamington Park is primarily subtropical rainforest and most treeholes occur among buttress roots. Mean reservoir surface area was 422cm$^2$ and they were 15.8cm deep on average. pH was around 5.9 and the mean conductivity was 226.5$\mu$S/cm. Leaf litter input was estimated to be 903g/m$^2$.

Table 7.1: The fauna of water-filled treeholes in Wytham Woods, England and Lamington National Park, Queensland (from Kitching, 1983).

| Taxa | WYTHAM WOODS | LAMINGTON |
|---|---|---|
| INSECTA | | |
| DIPTERA | | |
| Chironomidae | *Metriocnemus cavicola* | *Anatopynia pennipes* |
| Ceratopogonidae | *Dasyhelea dufouri* | *Culicoides angularis* |
| Culicidae | *Aedes geniculatus* | *Aedes candidoscutellum* |
| | *Anopheles plumbeus* | *Aedes* sp. |
| Syrphidae | *Myiatropa florea* | — |
| COLEOPTERA | | |
| Helodidae | *Prionocyphon serricornis* | *Prionocyphon* sp. |
| ARACHNIDA | | |
| ACARINA | | |
| ASTIGMATA | | |
| Hyadesiidae | — | New genus |
| PROSTIGMATA | | |
| Arrenuridae | — | *Arrenurus* sp. |
| MESOSTIGMATA | | |
| Ascidae | — | *Cheiroseius* sp. |
| AMPHIBIA | | |
| ANURA | | |
| Leptodactylidae | — | *Lechriodus fletcheri* |

The Wytham Woods treeholes supported a fauna consisting of one chironomid species, one ceratopogonid, two mosquitoes, a syrphid fly and a helodid beetle. The Lamington Park treeholes supported a fauna consisting of one chironomid species, one ceratopogonid, two mosquitoes, a helodid, three mites and a frog - in more than one group the genera present were the same as in Wytham Woods (Table 7.1). The two systems are thus very similar both faunistically and in terms of physical/chemical environment. Both have comparable trophic bases (largely saprophagic) but the Australian system has a substantial component of predators.

### 7.3 Pitcher plants:

As in treeholes, the dipterans are the most frequent inhabitants of the fluid-filled leaf chambers of pitcher plants. Fish and Hall (1978) recorded three species from *Sarracenia purpurea* which is common in glacial peat bogs in North America (Figure 7.3). These were, the flesh fly *Blaesoxipha fletcheri*, the mosquito *Wyeomyia smithii* and the midge *Metriocnemus knabi*. Fish and Hall found that only newly-opened pitchers attracted and captured insect prey and that these slowly decompose as the pitcher ages. This is accompanied by a lowering of the pH of the fluid. The relative abundance of the insect inhabitants of the pitcher is related to pitcher age as each of the three species consumes insect remains that are at different stages of decomposition. Specifically, the larvae of *Blaesoxipha* feed upon freshly-caught prey floating on the pitcher fluid surface, *Wyeomia* filter-feed on the partially decomposed material in the water column, and *Metriocnemus* feed on the remains that collect on the bottom of the pitcher chamber. A seasonal succession in the pitcher plant fauna is thus evident.

Much work has been done on the two species of nematoceran flies in *S. purpurea*. Kingsolver (1979) studied populations of *Wyeomyia smithii* in northern Michigan and found significant differences in larval development rate, the number of generations per year and larval mortality due primarily to microclimatic effects. He hypothesized that such mixed life history "strategies" would be favoured in the uncertain microaquatic environment of a pitcher plant. Lounibos *et al.* (1982) showed an interesting relationship between pitcher conditions and reproductive "strategies" in *W. smithii* over a wide geographic range. In southern U.S.A. populations, where density dependent constraints on larvae were deemed to be perpetually severe, the adult female mosquitoes required a blood meal in order for their eggs to mature. However, in northern populations, which

Figure 7.3: Photographs of the pitcher plant *Sarracenia purpurea* showing the entire plant, the pitchers and a longitudinal section of one pitcher with insect remains in its base.

exhibit periods of larval growth that are density independent (the mean number of larvae/pitcher decreases with increasing latitude), haematophagy is lost as reserves accumulated in the larval stages are sufficient for egg maturation in the adult.

Paterson and Cameron (1982) have shown a similarly complex life cycle for *Metriocnemus knabi.* Their data (Figure 7.4), based on a

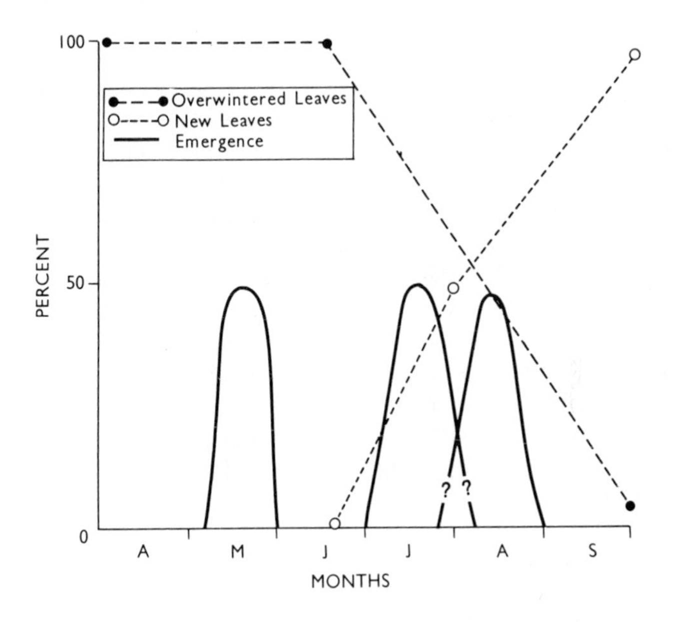


Figure 7.4: A preliminary model of the pattern of emergence of *Metriocnemius knabi* adults relative to the annual life cycle of the pitcher plant *Sarracenia purpurea* (redrawn from Paterson and Cameron, 1982).

study in New Brunswick, Canada, suggested that there were two cohorts in the population. Each cohort had three generations every two years, emerging in May and August of one year and July of the next year. One cohort thus produced two generations in even-numbered years and the other cohort produced two generations in odd-numbered years. Some intermixing of cohorts was possible during the midpoint of the July-August adult emergence period. This complex life cycle may be an evolutionary adaptation in this species or simply a pattern repeatedly reinforced by the seasonal characteristics of the host plant.

*Nepenthes* is a genus of pitcher plant that evolved in Indo-Malaysia and which supports communities of insects (primarily mosquitoes) and other organisms that live either in the pitcher fluid or on the upper walls of the chamber (Beaver, 1983). In a comparative study of the structure of the food webs within species of *Nepenthes* from different localities, Beaver (1985) found that complexity of the food web varied considerably. Species of *Nepenthes* in outlying areas such as the Seychelles, Madagascar and Sri Lanka had fewer trophic types of predator and prey, and fewer predator-prey interactions than pitchers of *N. ampullaria* and *N. albomarginata* in West Malaysia (West Malaysia > Madagascar > Sri Lanka > Seychelles; Figure 7.5). Communities in *Nepenthes* in outlying areas also had fewer and smaller guilds of species and more apparently empty niche space; however, the ratio of prey to predators remained fairly constant, around unity. Beaver postulated that the differences between the food webs were related, in a complex way, to the size of the country in which the particular species of *Nepenthes* was found, its degree of spatial and temporal isolation, the size of the local pool of species capable of living in pitchers, and the diversity of *Nepenthes* present. The maximum length of food chains within the most complex food webs was, however, probably limited by energetic constraints and/or environmental predictability.

## 7.4 Bromeliads:

Approximately 2,000 species of bromeliads are known from tropical and warm temperate regions of the Americas, and they range from altitudes of over 4,000m down to sea level. Tank bromeliads characteristically collect water in their leaf axils, a result of enlarged leaf bases that overlap tightly, and absorb nutrients from this water (Figure 7.6). Some bromeliads provide a relatively longterm, stable aquatic habitat for insects because the plants are long-lived; some may live for more than 20 years and achieve a volumetric capacity of 1.3 litres (Benzing, 1980).

Tank bromeliads represent perhaps the highest densities of phytotelmata found anywhere, presumably because of their epiphytic, three-dimensional distribution. In a cloud forest in Columbia, Sugden and Robins (1979) recorded a mean density of 17.5 plants/m$^2$ of ground area. Assuming the volume of water retained by each plant to be approximately 250cm$^3$, this suggests that the total amount of

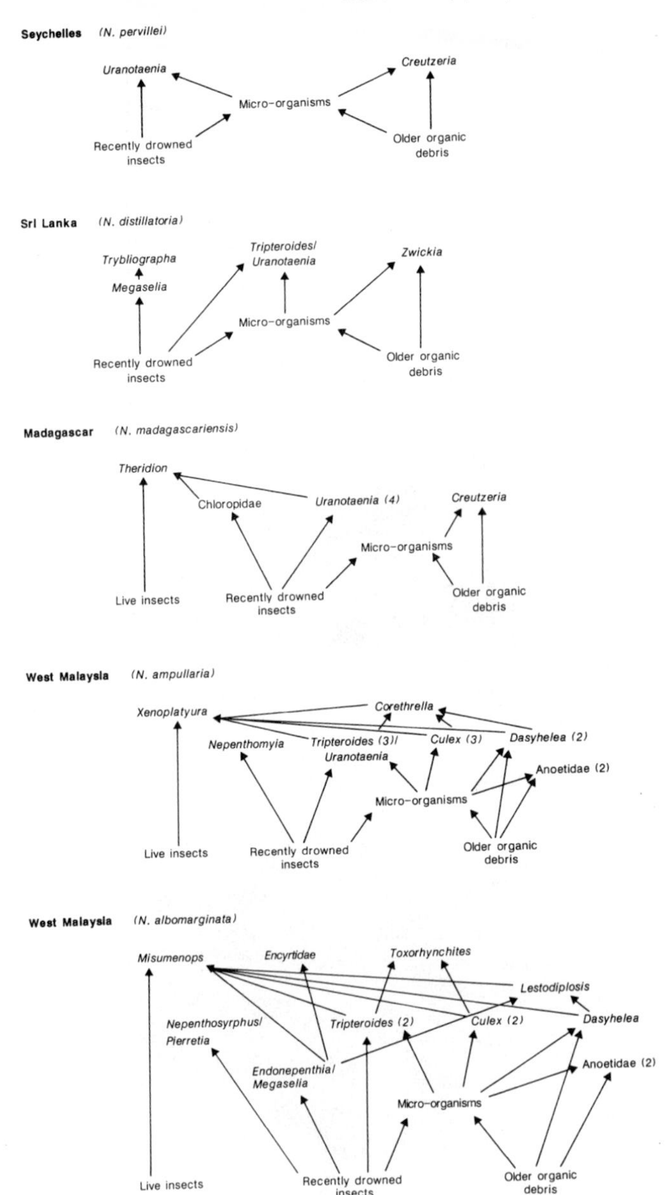


Figure 7.5: Foodwebs found in five species of the pitcher plant *Nepenthes* in four countries (numbers in brackets indicate more than one species present; redrawn from Beaver, 1983).

Figure 7.6: Examples of phytotelmata: (A) – an epiphytic bromeliad
(B) – a terricolous bromeliad

water available for colonization by aquatic animals in such a location would be more than 50,000 litres/hectare.

Bromeliads may be subdivided into two categories based on their method of nutrition: *Dendrophilous bromeliads* generally grow in forested areas and collect nutrients leached by rainfall from tree canopies. The tanks thus contain a nutrient-rich soup. *Anemophilous bromeliads* tend to grow on cacti in deserts or at the top of forest canopies. Here they obtain nutrients that are windborne.

Tank bromeliads exhibiting dendrophilous nutrition tend to have detritus-based food chains, while those with anemophilous nutrition support alga-based food chains (Frank, 1983). Food resources for insects in bromeliads are partitioned as the water is not held in one single cavity but in several smaller ones among individual leaf axils; migration between axils is known to occur in the larvae of some mosquitoes (Frank and Curtis, 1981).

Habitat selection is made by adult female insects and is dependent on a number of habitat features important to a particular species. Female mosquitoes, for example, select specific points on temperature and humidity gradients that occur from ground level to the top of the forest canopy. Shape of the plant, reservoir volume, condition of the leaves and presence of a flower spike also may be important factors.

Besides mosquitoes, other groups of dipteran insects commonly found in bromeliads include the chironomids, ceratopogonids, psychodids (moth flies), tipulids (craneflies), syrphids (flower flies), muscids, stratiomyids, phorids, tabanids, borborids, and anisopids. In addition may be found beetles (especially the Helodidae), bugs (Veliidae), one species each of stonefly (Plecoptera) and caddisfly (Trichoptera), many species of dragonfly (Odonata) and some frogs (Fish, 1983).

Corbet (1983) reported as many as 47 species (belonging to 17 genera) of dragonfly from phytotelmata, in general, and many of these live in both ground-dwelling or epiphytic bromeliads. There is insufficient evidence, however, to state their degree of specificity to these habitats. Phytotelmata seem to be more important for Andamselfly species (Zygoptera) than for dragonfly species proper (Anisoptera), as the former can be found in plants in the neotropical, Oriental and Pacific regions, and (though more rarely) in the afrotropical and Australasian regions. The latter are found, regularly, only in plants in the Oriental region and are represented by just two species of one genus. Odonate nymphs, because they can walk and climb, have wider options open to them than many of the other inhabitants of phytotelmata; *Roppaneura beckeri*, for example, can move between neighbouring plants of *Eryngium*.

It has been suggested that the Odonata may have evolved the habit of living in bromeliads through proximity of the latter to running water habitats (Calvert, 1911). The ancestral habitat of the Bromeliaceae is the riparian forests along the Amazon River. As we shall see in the next chapter, this river floods seasonally with the level of water fluctuating as much as 10m. At peak flood, the water level would be near the level of the lowest epiphytes. As bromeliads retain water for long periods of time, shallow-water, river species of dragonfly may have formed associations with the plants which may have become obligate in places or during times when rainfall was too patchy to maintain forest floor pools. Once established, such an association might well have allowed these species to extend their ranges into the forests beyond the influence of the river floods (Corbet, 1983).

## 7.5 Inflorescences:

Not only do leaf axils retain sufficient water to support aquatic communities but some large flower bracts do so as well. The genus *Heliconia*, for example, has some 150 species native to the Neotropical region but is also found in Indonesia, New Guinea and Samoa. The floral parts consist of a series of alternately-arranged, large bracts each of which surrounds 10 to 14 pairs of small flowers (Figure 7.7). The bracts of many species hold water (in *H. caribaea* the mean volume/bract is 8.4cm$^3$) which is derived from both rain and plant-transportation. Initially, the young flower buds are immersed in this water but as they mature they emerge and are held above the water (Machado-Allison *et al.*, 1983).

In *H. caribea* in the northern, lowland tropical rainforests of Venezuela, new bracts open approximately every 6.7 days and flowering occurs throughout the year although it peaks in the rainy season (June and July). The volume of the water reservoir increases with bract age as does the amount of organic carbon in the bract water. pH of the water is generally around 7.2. Machado-Allison *et al.* (1983) found larvae of 15 species of insects in *H. caribea*. Of the eight most abundant species, three are mosquitoes (*Wyeomia ulocoma, Culex pleuristriatus* and *Toxorhynchites haemorrhoidalis*), one a syrphid (*Quichuana* sp.), one a richardid (*Beebeomyia* sp.), and one a stratiomyid (*Merosargus* sp.). Of the non-dipterans, *Cephaloleia* and *Xenarescus* are chrysomelid beetles, the larvae of which feed by scraping the interior of the bract chamber. Densities of larvae of the three most abundant species, *Wyeomia, Culex* and *Quichuana*, correlated positively with the amount of rainfall. Predators in these bracts seem rare but intraspecific competition may occur. Adult females show distinct preferences for bracts of different ages when ovipositing. Bracts over ten weeks old may be less stable habitats than younger ones because they tend to be damaged by vertebrates. Food input to the community seems to be chiefly from decomposition of the flowers and the inner-wall of the bract.

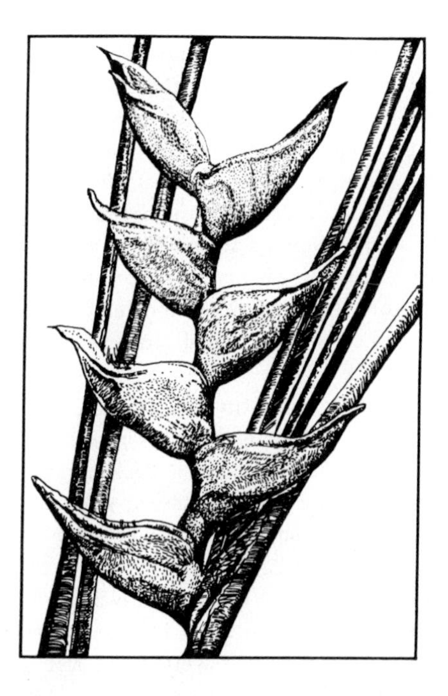

Figure 7.7: Inflorescence of *Heliconia caribaea* showing bracts and emerging flowers.

## 7.6 Snowfields:

A considerable number of organisms are associated with mountain snowfields. The biota can be subdivided into two basic groups: *(1)* those species that actually live on the surface of the snow during some stage in their life history, either finding their food or a mate there; *(2)* those species that are accidentally carried to high altitudes by air currents (perhaps during a dispersal phase in their life cycles) and which fall or settle onto the snow and subsequently die (Ashmole *et al.*, 1983). We shall concern ourselves only with the first group as it contains some forms which actively seek out snow as a moist environment.

Two prominent groups in the biota are the insects and the algae but enchytraeid worms (which are somewhat amphibious in nature) and mites are often present also. A number of insect species periodically migrate to snowfields to feed on the "animal fallout" discussed earlier, but these are adult, non-aquatic forms such as the muscid fly *Spilogona triangulifera* and the carabid beetle *Nebria gyllenhali* (Kaisila, 1952). By far the most numerous insects on snow are the Collembola (springtails) which appear to migrate up through the snow to the surface during thaw and back down to the soil as the temperature drops. Leinaas (1981a) found that the extent of migration was species dependent. For example, *Hypogastrura socialis* moves up and down in the snow profile according to temperature and is most abundant on the surface at air temperatures above $0^{o}C$; species of *Isotoma* show little vertical movement during winter and are mostly found in the coarse-grained bottom layer of snow but a few individuals appear on the snow surface during mild weather; a rich mix of species (e.g. *Entomobrya marginata* and *Lepidocyrtus lignorum*) tends to remain within the snow pack particularly in areas of coarse-grained snow. As the guts of many species of Collembola have been found to contain only small amounts of complex plant residues, which are more abundant in the underlying soil, Leinaas suggested that these insects do not move into this nival habitat to seek food. They may be moving up into the snow in order to avoid compacting ice on the soil surface during mid winter and also water-logging of the soil as the snow melts in late winter. Both Leinaas and Zettel (1984) have suggested that snow-surface activity may be important in terms of the dispersal of populations of Collembola. Many species are hygrophilic and so dispersal, under the dry conditions of summer, may be impossible. By moving across snow surfaces, in winter, the springtails can cover considerable distances under optimal humidity.

As a physical adaptation to jumping on snow, Leinaas (1981b) has shown that a cyclomorphosis occurs in *Hypogastrura socialis*. In summer, the teeth on the furcula (abdominal jumping organ) are small, but in winter the teeth are larger and the surrounding cuticle becomes thickened and this may improve the insect's grip on the snow surface. Some winter-active insects are protected from freezing when walking on snow by the presence of antifreeze agents in their bodyfluid which cause a thermal hysteresis between the freezing-melting curves thus preventing the growth of ice crystals within the temperature range at which the insects are active. In the snow scorpionfly (*Boreus westwoodi*; Mecoptera), for example, there is a difference of 5 to 6°C between the melting point and the temperature of ice growth in the haemolymph indicating the presence of antifreeze compounds. These compounds are likely to be proteins or glycoproteins and, in *B. westwoodi* they prevent the growth (not the formation) of ice crystals down to -7°C which covers the minimum temperature at which they are active (-2°C) with a considerable margin of safety (Husby and Zachariassen, 1980). There do not appear to be comparative values for the Collembola.

Craneflies (Tipulidae) also occur in snow and adults of three species of *Chionea* have been found in the free air space under snow in Finland. The larvae are thought to live in tunnels inside damp grass hummocks where they may be protected from severe frosts (Itamies and Lindgren, 1985).

Considerable numbers of algae occur in snow where they frequently give a reddish tinge to the surface layers. They most commonly belong to the families Chlorophyceae (greens), Cyanophyceae (blue-greens) and Xanthophyceae (yellow-greens) but diatoms (Bacillarophyceae) may be present too. Gram-negative bacteria and microfungi may form associations with some types of algae (Lichti-Federovich, 1980).

*Chlamydomonas nivalis* is a unicellular, motile, green algae found commonly in snow fields in many parts of the world (e.g. U.S.A., Greenland, and Australia). The most common form found in snow is a resting stage which typically exhibits a red colour due to the presence of carotenoids in cytoplasmic vacuoles (Marchant, 1982); the colour may extend to depths of up to 10cm into the snow. These resting cells are roughly spherical, 10-50$\mu$m in diameter and have a thick cell wall with a smooth outer surface. On the outside of the wall is a loose network of fibres that intermesh with encapsulated bacteria and surface debris. The bacteria, too, have a thick capsule wall and it is thought that this and the thick algal cell wall are important structures in protecting the cells from dehydration and freeze-thaw cycles as the snow melts (Weiss, 1983). Hoham *et al.* (1983) have shown that a succession of species of the alga *Chloromonas* occurs and that this may be due to different tolerance levels of individual species to environmental factors such as light intensity, carbon dioxide concentration and temperature, factors that vary according to the age of a snowfield.

## 7.7 Deserts:

Two groups of algae occur in deserts, Edaphic (or soil) algae and Lithophytic (or rock) algae. Friedmann *et al.* (1967) subdivided edaphic algae into three categories: *(1) Endedaphic algae* live in soil and occur in many deserts throughout the world. Frequently they occur to depths of greater than 50cm where they may be washed by rain and where they subsist at reduced metabolism for long periods (Schwabe, 1963). Blue-green algae cultured in the laboratory from the Atacama Desert of northern Chile are slow to germinate and grow but are very resistant to desiccation and high levels of irradiation. These algae grow to form small convex, cushion-like pellets of sand grains cemented together by the gelatinous sheaths of their filaments. It is thought that these structures shade the small area of soil beneath them thus preventing evaporation and thus creating a minute water reservoir (Forest and Weston, 1966). *(2) Epedaphic algae* live on the soil surface and are also widely distributed in deserts and semi-deserts throughout the world. In deserts of the southwestern United States, Cameron and Blank (1966) found three forms: *(a) Raincrusts* - which arise in shallow depressions in the soil where rainwater collects. They typically consist of a thin layer of algae that warps, curves upwards and finally breaks into fragments when dry. *(b) Algal soil crusts* - which are soil-like in colour and seem to be more stable in structure, perhaps due to the presence of filamentous blue-green algae. *(c) Lichen soil crusts* - which are the largest and thickest (up to 2-5cm) of the formations. Here, fungal hyphae bind the algae into a lichen association and the resulting mass frequently becomes folded and sticks up from the soil surface. *(3) Hypolithic algae* live on the under surfaces of stones and are found in both hot and cold deserts. The stones are typically translucent, to some extent, thus light reaches the cells at intensities of between 30.0 and 0.06% of surface illumination (Vogel, 1955). Because rainwater penetrates the soil most easily along the stone-soil boundary, small reservoirs of water persist under these rocks even after it has evaporated from the surrounding soil. The hypolithic community typically consists of filamentous and coccoid blue-green algae and green algae (Friedmann *et al.*, 1967).

Lithophytic algae live in, rather than on, rocks. They are of two types: *(1) Chasmolithic algae* live in the fissures of rocks. *(2) Endolithic algae* live within the rock matrix itself, colonizing spaces between particles. Again, the rocks tend to be somewhat translucent and porous. The latter feature is important in trapping and retaining moisture by capillary force so as to provide the necessary water balance for the algae (Friedmann, 1964).

Provision of water to all these algal associations may be through rain, although in many cases it is derived from condensation of dew at night. Despite this, the algae are sometimes subjected to total loss of environmental water in which case they are able to survive through physiological tolerance. Parker *et al.* (1969), for example, have shown some soil algae to survive drying for over 60 years; Cameron *et al.* (1970) determined that some filamentous and coccoid blue-green algae could survive continuous high vacuum for 5 years; and Trainor (1962) showed that the green alga, *Chlorella*, can survive desiccation at 130°C for 1 hour.

## 7.8 Dung:

Stretching our already broad definition of temporary aquatic habitats to near breaking point, we should perhaps include the communities found in animal dung. Cow dung produced in summer, when the animals are grazing on fresh grass, is quite liquid and has a high water content (73-89%; Hammer, 1941). The fauna consists mostly of fly larvae, from a variety of families, and also beetles. Laurence (1954) studied cow pats in Hertfordshire and revealed a diverse and abundant fauna (Table 7.2). Support for regarding these habitats as semiaquatic ones is evident from the similarity between some of the taxa found in them and in the summer terrestrial fauna of temporary ponds (Chapter 4; Figure 4.9). Taxa common to both habitats are larvae of the dipteran families Sphaeroceridae, Tipulidae, Sepsidae, Ceratopogonidae and Chironomidae, as well as beetles belonging to the families Hydrophilidae (scavenging waterbeetles) and Staphylinidae (rove beetles).

The habitat itself is a changing one depending on intrinsic properties such as aging of the material itself, and extrinsic factors such as weather. After deposition, the pat begins to cool down to air temperature, and, in summer, a crust forms on the surface, usually within 24 hours. This crust separates from the more fluid dung underneath

Table 7.2: Maximum number of larvae found in one 2.5 cm core of cow dung (after Laurence, 1954).

| Family | Common name | Genus | Date | Number |
|---|---|---|---|---|
| PSYCHODIDAE | (moth flies) | *Psychoda* | 17 Oct. | 598 |
| SPHAEROCERIDAE | (dung flies) | *Limosina* | 1 Apr. | 230 |
| SEPSIDAE | (scavenger flies) | | 18 June | 76 |
| SPHAEROCERIDAE | (dung flies) | *Copromyza* | 27 Jan. | 56 |
| TRICHOCERIDAE | (winter gnats) | *Trichocera* | 5 Dec. | 37 |
| SCATOPHAGIDAE | (dung flies) | *Scatophaga* | 1 Nov. | 53 |
| MUSCIDAE | (predaceous flies) | | 17 Oct.,30 July | 7 |
| HYDROPHILIDAE | (scavenging waterbeetles) | | 15 June | 7 |
| SCATOPSIDAE | (dung flies) | *Scatopse* | 13 June | 10 |
| ANISOPODIDAE | (window flies) | *Anisopus* | 23 July | 29 |
| CERATOPOGONIDAE | (biting midges) | *Culicoides* | 19 May | 58 |
| MUSCIDAE | (flies) | *Polietes* | 17 Oct. | 7 |
| MUSCIDAE | (flies) | *Dasyphora* | 7 May | 14 |
| CALLIPHORIDAE | (flesh flies) | *Mesembrina* | several | 1 |
| EMPIDIDAE | (dance flies) | | 19 Sept. | 9 |
| STRATIOMYIDAE | (soldier flies) | | 18 June | 19 |
| CHIRONOMIDAE | (midge) | *Smittia* | 14 Nov. | 15 |
| STAPHYLINIDAE | (rove beetles) | | 10 July | 37 |

and an air space is formed between the two. The crust appears to seal off the odour of the dung and the pat becomes less attractive to many adult flies. Much of the egg-laying must therefore occur soon after deposition. Winter-produced dung is coarser in texture, due to a change in cattle fodder, and tends not to crust over.

The fly larvae that inhabit cow dung show many respiratory features in common with aquatic and semiaquatic species and can be broadly categorized into: *amphipneustic* (spiracles on the prothorax and terminal segments) - many species; *apneustic* (no spiracles) - chironomids and ceratopogonids; *metapneustic* (spiracles only on the terminal segment) - tipulids (Keilin, 1944). An interesting seasonal change noted by Laurence (1954) was that larvae of species living in the less-aqueous, winter dung do not show the modification of the posterior end (i.e. spiracles borne on the end of a long, telescopic siphon) seen in some summer species.

In the Hertfordshire dung, a clear seasonal succession of species was evident (Figure 7.8) and few genera were present

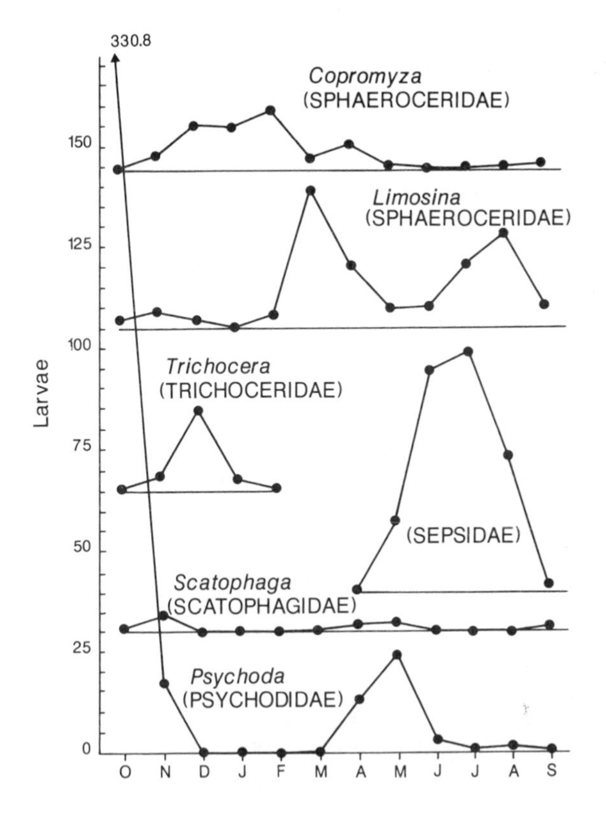


Figure 7.8: Seasonal succession of larval insects in cow pats (numbers of larvae are recorded as geometric mean number of larvae per pat sample of four 2.5 cm cores for each month; redrawn from Laurence, 1954).

throughout the year. The genus *Limosina* (dung flies) is exceptional in that several species succeed one another throughout the year. Larvae of the Muscidae and Sepsidae (scavenger flies) were absent in winter, while larvae of *Trichocera* (winter gnats) and also *Smittia* (chironomids; not shown) occurred mainly in the winter and entered a diapause during the warm summer months. Larvae of *Copromyza* (dung flies) occurred primarily in winter and spring, while larvae of other dung flies, *Scatophaga*, had two peaks, one in April-May, the other in September-December. Larvae of *Psychoda phalaenoides* have a very rapid rate of development and were very abundant in the late autumn with another, lesser, peak in late spring.

The rate of growth in some species is very high, some coprophagous muscid larvae, for example, mature in just 5 days in summer, and most summer-dipteran species have a development period of less than 40 days. In some of the stratiomyids (soldier flies) the period of development is long (60-100 days), as is the case for the chironomid *Camptocladius*. For species such as these, the pat probably disintegrates before development is complete. Periods of development may vary between species of the same genus as, for example is seen in the scavenger flies *Sepsis* (Figure 7.9).

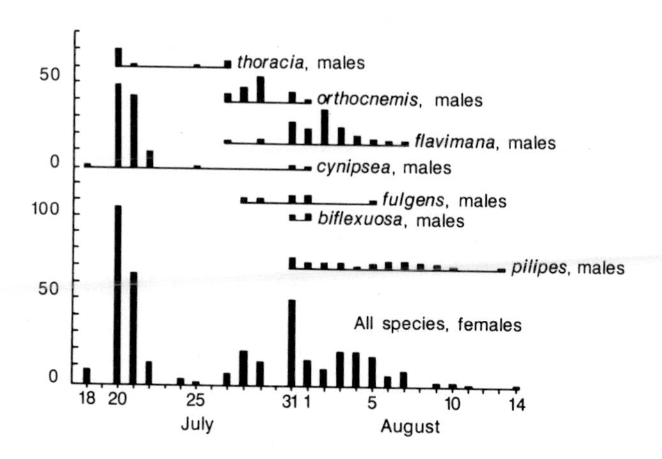


Figure 7.9: Emergence succession of species of *Sepsis* from dung deposited on 5 July, 1950 (redrawn from Laurence, 1954).

How numerous are these habitats and how significant are their faunas? Pat size varies between 13 and 45cm in diameter, and a cow may deposit at least six per day. A single bullock may produce 26kg of dung each day per 500kg of body weight, or roughly nineteen times its own weight in one year (Henry and Morrison, 1923). Approximately one-eightieth of the weight of the dung is attributable to the animals living in it and so, in one year, a cow will leave in its dung enough material to support an insect population equal to one fifth of its own weight (Laurence, 1954).

## 7.9 Saltwater rockpools:

A final type of temporary freshwater habitat is represented by the brackishwater pools of the marine supra-littoral zone. Because of their intermittent position between land and sea, the salt content, and therefore their freshwater content, often fluctuates widely. Ganning

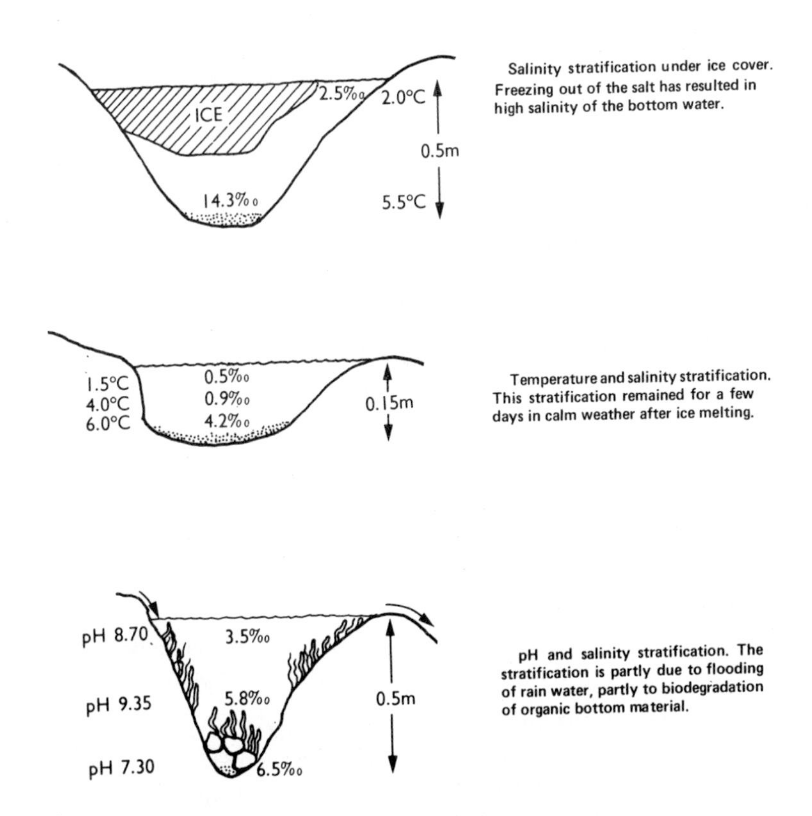


Figure 7.10: Changes in salinity of rockpools due to various factors (re-drawn from Ganning, 1971).

(1971) measured salinity changes in Swedish rockpools and correlated these with environmental changes. Figure 7.10 shows some of the

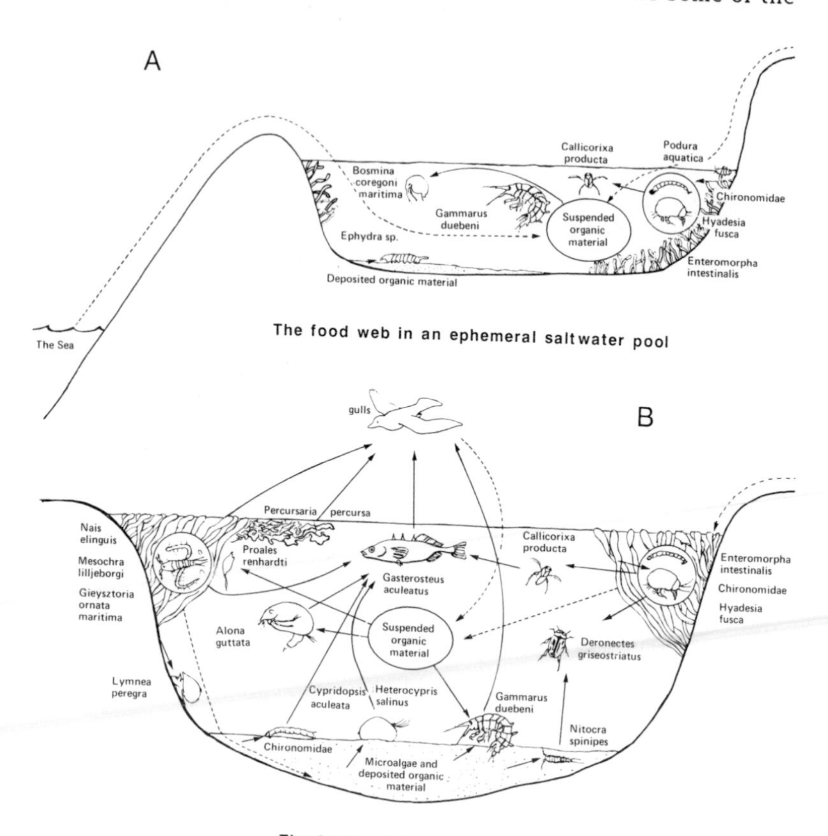


Figure 7.11: Examples of foodwebs in rockpools in the Baltic area (redrawn from Ganning, 1971).

factors that contributed to these changes. He produced a classification scheme of rockpools based largely on salinity levels but warned that these pools are not static and stable biotopes, but dynamic and changing, and consequently defy absolute categorization. His classification scheme spans permanent saltwater pools to permanent freshwater pools, however only two of his pool types are particularly relevant to our discussion. These are the ephemeral saltwater pool and the brackishwater pool. Ephemeral saltwater pools are generally of small volume and are situated high up on the shore. They receive saltwater input from splashing during strong winds at high tide. Physical and chemical parameters of the habitat vary greatly and the biota is limited. Figure 7.11A shows some of the characteristics of the community living in these pools in the Baltic. Brackishwater pools are supplied by both rain and saltwater spray but are generally lower on the shore. The distinction between these pools and the ephemeral saltwater pools is based chiefly on size, amount of bottom sediment, degree of salinity and, though to a lesser extent, the presence of the filamentous alga *Enteromorpha intestinalis* (it is more abundant in brackishwater pools). Figure 7.11B indicates that the biota is somewhat richer in these latter pools.

In a study of the fluctuations in salinity in salt marsh pools in Northumberland, England, Sutcliffe (1961) found a gradation in salinity. Some pools were flooded only by seawater at very high spring tides and their salinity was generally less than $15^{o}$/oo. Other pools were flooded more frequently and their salinity was generally greater than $15^{o}$/oo. A few species of euryhaline insects bred regularly in these pools, namely a caddisfly, a beetle, two species of midge, a mosquito and a shore fly. In addition, during the summer, there were irregular influxes of adults of various species of bugs and beetles, however these only colonized pools with salinities less than $10^{o}$/oo.

In a detailed study of the faunas of 14 rockpools along the north shore of the St. Lawrence River, in Canada, Williams and Williams (1976) found certain assemblages of animals to be characteristic of low and high salinities. Figure 7.12 shows that most of the insects were confined to lower salinities, for example, the dragonfly *Aeshna interrupta*, the caddisflies *Limnephilus tarsalis* and *Oecetis*, and the beetles. A few insect species, however, such as the shore fly *Ephydra subopaca* and the marine midges *Halocladius* and *Cricotopus sylvestris*, were most abundant at salinities approaching full-strength seawater. Survival in high salinities requires a high degree of osmotic independence that may only be achievable by having, amongst other things, a body that tends to be impermeable to water and salts (Sutcliffe, 1960). It has been suggested that osmotic regulation and the problems of respiring in saltwater (the ideal respiratory membrane is thin, of large surface area and semipermeable) may involve the evolution of such different physiological adaptations that few insects have been successful in doing so and thus have been prevented from widespread invasion of marine environments. Consequently,

Figure 7.12: Relationship between salinity and the occurrence of various animal species in 14 rockpools on the north shore of the St. Lawrence River, Canada (redrawn from Williams and Williams, 1976).

"marine" insects are, for the most part, still restricted to "bridging" habitats between land and sea (Cheng, 1976). Saltwater rockpools may therefore be the crucibles in which the evolution of improved physiological adaptations are taking place. This may apply to inland saline lakes, ponds and estuaries also.

### 7.10 Starfish:

One of the most unusual temporary marine habitats for an insect is seen in the caddisfly *Philanisus plebeius.* In New Zealand, eggs of this species develop in the coelomic cavity of the starfish *Patiriella regularis* (Winterbourn and Anderson, 1980). Larvae migrate from the host soon after hatching and then are freeliving in the intertidal zone where they live amongst attached algae and feed primarily on red algae (Rhodophyceae). It is not known exactly how the eggs are deposited in the starfish or how the larvae leave but it is hypothesized that oviposition occurs through the papular pores (abundant on the aboral surface of the starfish) and that newly hatched larvae may leave by this same route or through the wall of the stomach.

# 8 APPLIED ASPECTS OF TEMPORARY WATERS

## 8.1 Aquaculture/agriculture rotation - an ancient art:

A classic example of the large scale, applied use of temporary waters (see Mozley, 1944) is to be seen in the Dombes ponds of France. The Dombes is a platform to the north-east of Lyon consisting of long morainic mounds in fan-formation created by Quaternary glaciers. The moraine itself (approximately 280m high) was formed during the Riss glaciation. During the late Wurm glaciation, substantial amounts of loess (wind-borne, calcareous silt, 0.002-0.05mm in grain diameter) were deposited in the depressions between the morainic mounds (Figure 8.1). Post-glacial rains leached much of the loess creating decalcified clayey soil which provided a waterproof lining to these basins.

Up until the thirteenth century, only a few natural ponds ("leschères") existed but these were stocked with fish. During the Middle Ages, because of a demand for fish (for religious reasons) and the fact that the soil in the region was poor (clay) yet retained water, large numbers of ponds were created. Making a pond simply entailed damming the ends of the depression between morainic mounds. This caused social problems however, as the land so flooded frequently belonged to several landowners. To alleviate this, "The Rules of Villars" ("La Coutume de Villars") were proposed and have governed the use of the ponds from the fourteenth century until today (Truchelut, 1982).

The rules proposed a rotation of use of the ponds: two years full, one year dry (Figure 8.2). Water for filling a pond ("evolage") generally comes either from a pond situated at higher elevation or from a system of ditches which lead into the pond and which collect rainwater. Water level in the pond is controlled by a sluice-gate ("thou"). When filled, each pond is then stocked with fish, usually carp. After two years of "evolage" the fish are harvested and marketed, the profits going to the collective owners of the evolage. As soon as the pond bed is dry enough to be workable, it is ploughed (except for the central channels and the fishing area which are never drained) and planted with cereals ("assec"). The grain is harvested at the end of the summer and the profits distributed among the owners of the assec - who may be different from the owners of the evolage. Stubble

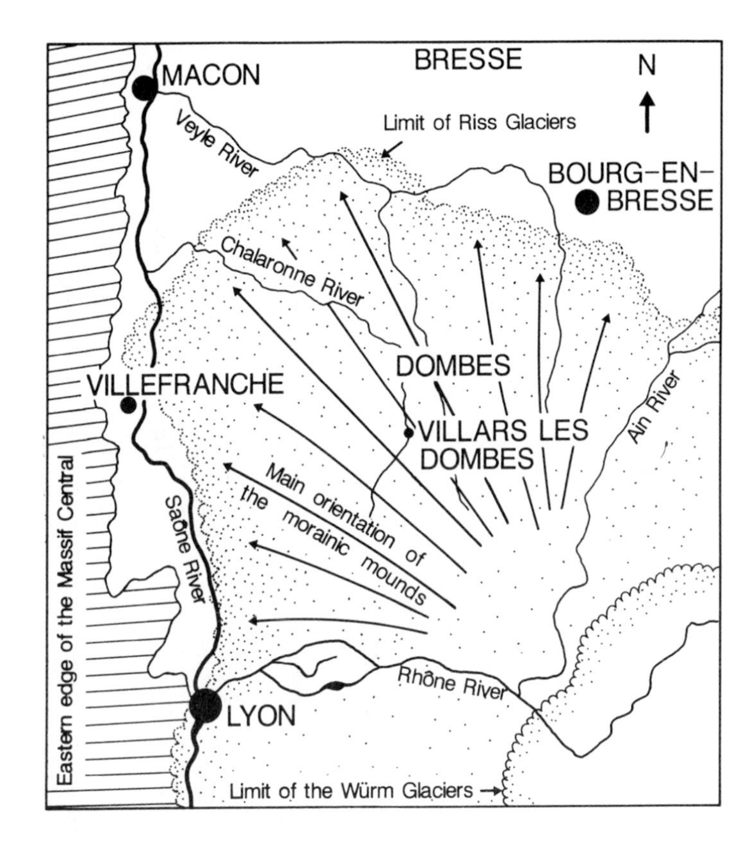


Figure 8.1: The origins of the Dombes ponds, France (after Tachet).

and roots left behind form a source of food for aquatic organisms when the pond is reflooded at the start of a new evolage. Because of the assec/evolage time ratio, and the fact that the ponds are often arranged in a series, approximately one third of the Dombes ponds are dry in any one year and two thirds are flooded.

This rotational system has changed little since the fourteenth century apart from a reduction in the total surface area of the ponds (19,000 hectares in 1862 to 11,000 hectares at present) and occasional change in the relative duration of the evolage/assec stages (e.g. to 3:2 or 4:2).

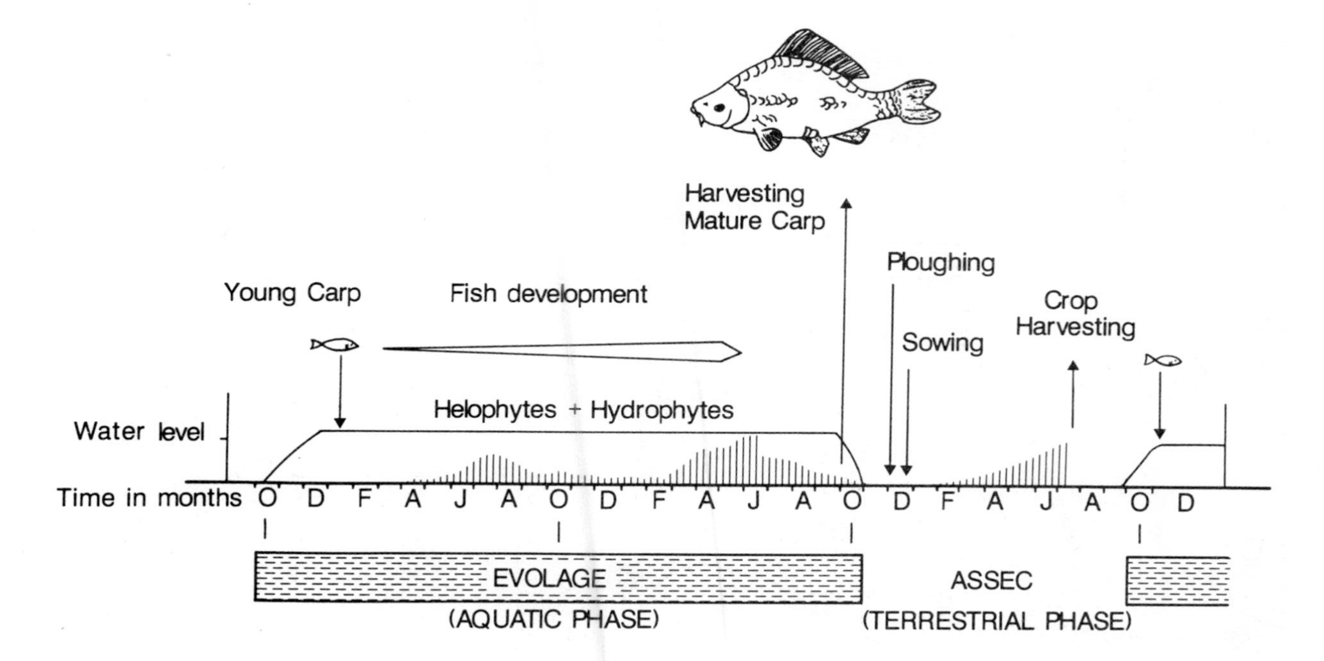


Figure 8.2: The traditional "Evolage-Assec" rotation in the Dombes ponds (after Tachet).

Currently, the Dombes region supports over 1,000 ponds each of surface area somewhat less than 10 hectares. Each pond is formed behind a dyke, made from compacted soil, and a channel runs through its middle and widens on the "upstream" side - this is the fishing area ("pêcherie") (Figure 8.3). The pêcherie and channel may be as much as

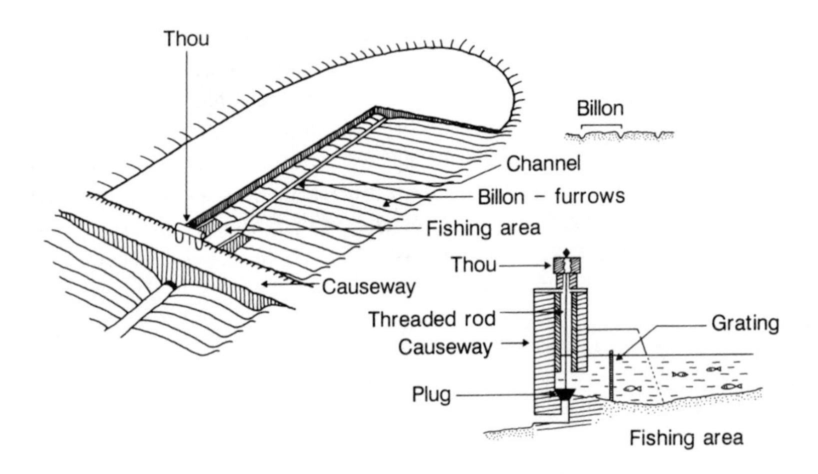


Figure 8.3: (A) Simplified structure of a Dombes pond; (B) Section through a "Thou" (after Tachet).

2m deep but the bulk of the pond is between 0.5 and 0.8m. Flow of water through the dyke is controlled by a conical plug connected via a threaded rod to the "thou", the sluice-gate mechanism. A grating prevents fish from escaping through an overflow in times of high water. To assist drainage during the assec, furrows are ploughed at right-angles to the channel with a unique pattern of 10-12 shallow furrows to one deep one (the "billon" structure).

Fish are usually harvested between September and April, and apart from the large ponds, there are smaller ones which contain the breeding stock. Mirror carp (*Cyprinus carpio*) is the predominant species cultivated (65% of the stock), followed by roach (*Rutilus rutilus*; 20%), tench (*Tinca tinca*; 10%) and pike (*Esox lucius*, 5%).

Biologically, the Dombes system is on firm ground as it improves soil of basically poor quality. The soil is allowed to lie fallow, under water, while fish are being raised. Aquatic plant debris from the evolage is mineralized when exposed to the air as the ponds are drained - if left submerged (in a non-degradable, acidic environment) much of this debris would accumulate on the pond bottom and its nutrients would not be released. However, Grant and Seegers (1985) have recently shown that in non-fertilized paddy fields, lowland rice uses ammonium-nitrogen present at the time of flooding and also that nitrogen mineralized from soil organic matter. Experiments using a species of *Limnodrilus* taken from Philippine rice soil showed that these oligochaete worms readily increased the ammonium-nitrogen production, probably by promoting the breakdown of soil organic matter through ingestion of soil and excretion of compounds in various stages of mineralization. These nutrients thus become available to the terrestrial crops in the assec and, in turn, their remains provide nutrients for the base of the aquatic food chain (algae, macrophytes, invertebrates) upon which the fish depend.

## 8.2 Floodplain fisheries:

### *8.2.1 Habitat types*

In areas which are subject to large seasonal changes in rainfall (e.g. monsoon regions of the tropics), many rivers cyclically overflow their banks. These dramatic fluctuations in water level result in local periodic episodes of flood and drought that affect not only the river itself but also tributaries and ponds on its floodplain. Associated with these cycles are significant changes in water chemistry and primary and secondary production. The aquatic floras and faunas of these floodplain waterbodies are thus subject to spatial and temporal fluctuations in their environment of the nature of those discussed in Chapter 3. Despite this, some floodplain waters have diverse and abundant faunas, especially fishes, although these have not been studied in any great depth (Welcomme, 1979).

Because of the individual nature of the topography and flow characteristics of any one river, the waterbodies on the floodplain vary both qualitatively and quantitatively between systems. In addition, current-induced shifts in sediment frequently change the nature of the streams and pools within single systems. Figure 8.4 illustrates the features of geomorphology likely to be encountered on a tropical floodplain. They are:

1. The river channel.

2. Oxbows or oxbow lakes, representing the cutoff portions of meander bends.

3. Point bars, loci of deposition of the convex side of river curves.

4. Meander- scrolls, depressions and rises on the convex side of bends formed as the channel migrated laterally downvalley and toward the concave bank.

5. Sloughs, areas of dead water, formed both in meander-scroll depressions and along the valley walls as flood flows move directly downvalley, scouring adjacent to the valley walls.

6. Natural levees, raised berms or crests above the flood plain surface adjacent to the channel, usually containing coarser materials deposited as flood flows over the top of the channel banks. These are most frequently found at the concave banks. Where most of the load in transit is fine-grained, natural levees may be absent or nearly imperceptible.

7. Backswamp deposits, overbank deposits of finer sediments deposited in slack water ponded between the natural levees and the valley wall or terrace riser.

8. Sand splays, deposits of flood debris usually of coarser sand particles in the form of splays or scattered debris. (Leopold *et al.*, 1964)

The main channel and its branches (A and B respectively on Figure 8.4) are usually permanently water-filled, although they do not always flow year round. Ancilliary to these are channels and streams that flow across the levees and connect the main channel with lentic waterbodies on the floodplain. Many of these lesser channels are intermittent but they are important routes for seasonally-migrating fish. The lentic habitats are those left after the main channel has overflowed and returned to its normal level. Many are temporary in nature and they include sloughs in oxbows (C), meander-scroll depressions (D), backswamps (E) or channels left by a previous course of the river (F). During the peak of the flood, these waterbodies are frequently temporarily merged into one (Welcomme, 1979).

Most of the major seasonal floodplains occur in the tropics and subtropics and, in fact, most tropical rivers have fringing, internal or coastal deltaic floodplains somewhere along their course, e.g. the San Antonio River, the Orinoco River and the Amazon River (Central and South America); the Senegal, Niger, Zambezi, Nile and Volta rivers (Africa); and the Euphrates, Tigris, Indus, Ganges and Mekong rivers (Asia). There are, however, floodplain rivers in temperate regions of the world, e.g. the Danube and Amur (Asia), although many of these

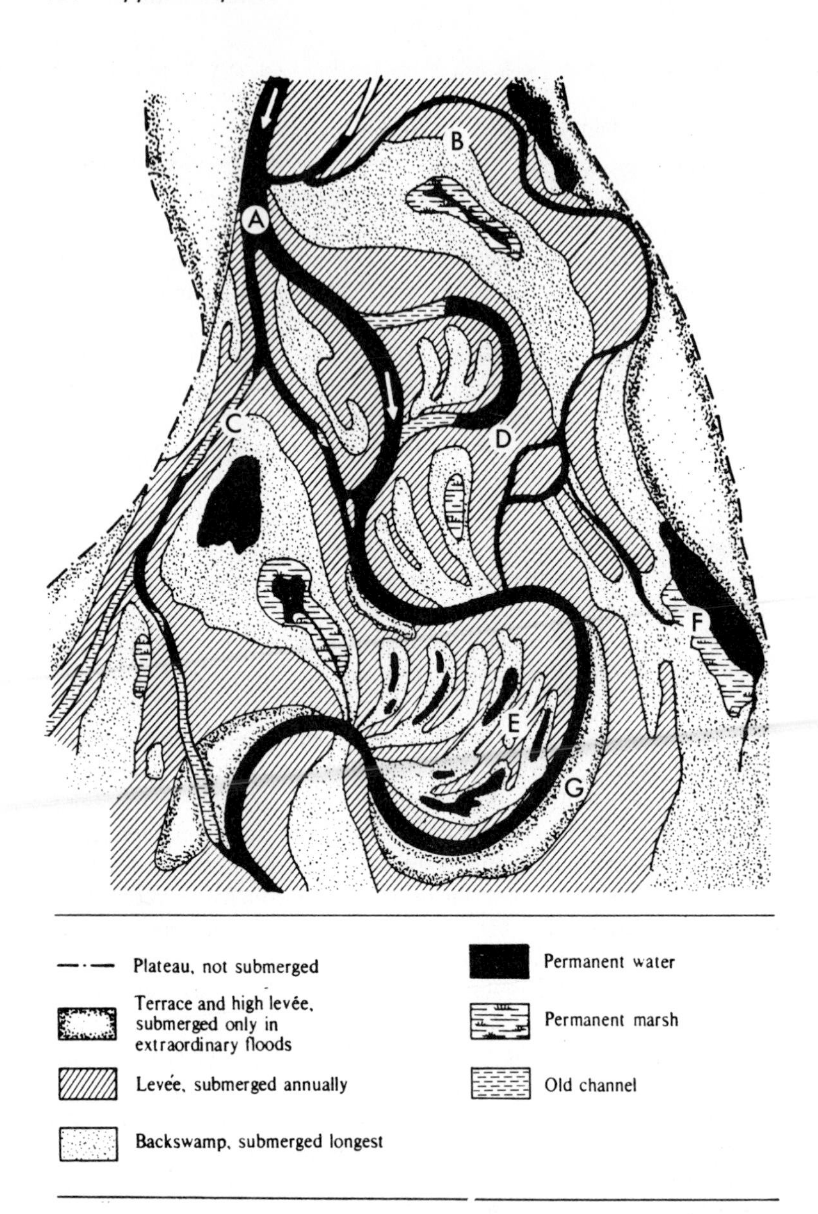


Figure 8.4: Diagram of the main geomorphological features of a tropical floodplain (letters refer to description in the text; redrawn from Welcomme, 1979).

have been altered by flood-control programmes. Perhaps the largest floodplain system in the world is the "Pantanal" (area approximately 93,000km$^2$) of the Rio Paraguay in the border area of Bolivia and Brazil. Here the annual flood cycle temporarily transforms savanna into a vast wetland. Unfortunately, little seems known of the ecology of this area.

In the tropics, basins on the floodplain lose water, through evaporation and seepage, throughout the dry season; thus they tend to diminish in size with increase in time since the last flood. This has the effect of concentrating the nutrients dissolved in the water which is an important process in regions like the Amazon where the levels of nutrients in the main river channels are extremely low. Productivity in the floodplain lakes is thus much higher than in the river. In Amazonia, these Várzea lakes (Figure 8.5) are filled annually and are

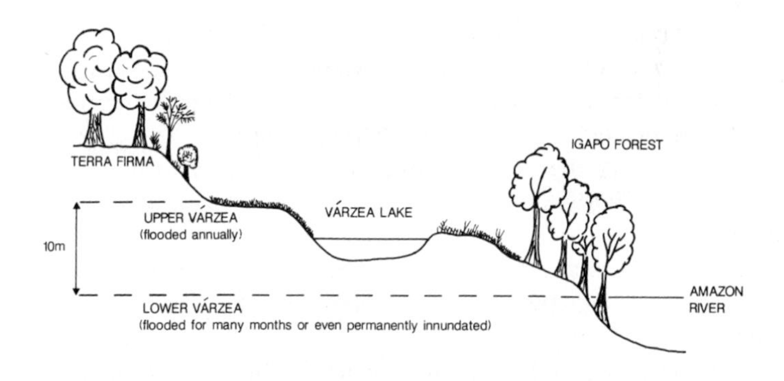


Figure 8.5: Position of a Várzea lake on a floodplain in central Amazonia.

situated on the floodplain between higher ground (terra firma), which is never flooded, and the lower igapo forest which is an area subject to prolonged or even permanent submersion. In central Amazonia, the vertical height between the dry season level of the water in the river (=level of the igapo forest) and the wet season, high water mark (=level of the upper Várzea) is as much as 10m.

### 8.2.2 *The flora*

Phytoplankton biomass increases in floodplain waterbodies as they evaporate and reaches a peak in the dry season. This is quite simply due to the concentrating effect of the nutrients. However, once nutrients have been depleted, the populations crash and production is low until the next flood. In many regions, phytoplankton production is limited to the surface layers because of colouring of the water which limits light penetration. In western Amazonia, for example, waters derived from the Andes are loaded with suspended particles derived from basic rocks and have a cloudy-white appearance. In central Amazonia, waters are high in humic substances and are consequently dark-brown or even black, e.g. as in the Rio Negro watershed.

Epiphytic algae may be more important, in terms of production, than phytoplankton (Ducharme, 1975). Most of the primary production in the floodplain waters is derived from macrophytes. Their role is in:

(A) providing diverse habitats for plants and animals;

(B) acting as a filter and trap for allochthonous and autochthonous materials which are nutrients not only for the plants themselves but also for associated aufwuchs, invertebrates and fish;

(C) increasing the concentration of elements in littoral areas of lakes and in newly-flooded waters of the floodplain, via a nutrient pump effect;

(D) contributing to autotrophic production, through decay, by forming a rich detritus which is used as food by a variety of organisms;

(E) functioning as a nutrient sink which, by locking up nutrients during the flood phase and releasing them to the soil during the dry season (through decay, burning and subsequent production of ash, and by being grazed and returned as dung), results in the conservation/retention of nutrients in the floodplain, rather than their being carried downstream (Howard-Williams and Lenton, 1975; Welcomme, 1979).

These macrophytes include a variety of grasses (primarily on savanna-type floodplains), which are mostly rhizomatous and flood-resistant, and dense forests of flood-resistant trees (in the rainforest regions of Africa, South America and Asia, e.g. the igapo forests of Amazonia). Other important elements in the flora are the floating forms, both free-floating species and those that form floating mats or meadows. Many of the free-floating species have cosmopolitan distributions, for example, water hyacinth (*Eichhornia crassipes*), water lettuce (*Pistia stratiotes*), water velvet (*Azolla* sp.) and salvinia (*Salvinia* sp.). Part of the success of these plants in these waters is due to their ability to rise up and down with the water level thereby remaining in optimal light conditions. Many submerged plants are forced to become dormant for that part of the year when they are covered by turbid waters. Growth of floating forms is so high that the water hyacinth, for example, can double in size every 8-10 days given warm, nutrient-rich water (Wolverton and McDonald, 1976). The long, flowing, densely-branched root clusters of many of these species (designed to absorb the few nutrients from the water) provide sites of attachment for invertebrates.

### *8.2.3 The fauna*

Zooplankton is more abundant in the waters of the floodplain than in the main channel, as flowing water does not suit them. Here, their numbers are controlled by factors such as phytoplankton biomass (food), water temperature regime, length and severity of drought, availability of oxygen, turbidity, and amount of vegetation. Moghraby (1977) has shown that in many species that live in floodplain pools alongside the Blue Nile, adults or eggs enter a diapause, as water temperature drops and turbidity increases at the onset of a flood, and survive in the bottom sediments until temperature and turbidity return to dry season levels.

Little is known of the benthos of the meandering, slow-flowing, silty floodplain rivers, but diversity and abundance are thought to be low (Monakov, 1969) although this may vary according to substrate composition (Blanc *et al.*, 1955) as in other benthic habitats. The benthos of permanent lakes on the floodplain seems dominated by oligochaetes, molluscs and chironomids and biomass is extremely variable (e.g. low in profundal areas but often high in sheltered, shallow bays), due perhaps to lack of oxygen in deeper waters and periodic mass emergence of some of the insect species (Rzoska, 1974).

Temporary ponds and lakes on the floodplains, again, have been little studied but molluscs, particularly pulmonate snails, appear to be prominent in the benthos (Blanc *et al.*, 1955; Welcomme, 1979) although their diversity may be low (Bonetto *et al.*, 1969). Ephemeroptera, Trichoptera, Chironomidae, Hemiptera and Mollusca were abundant and widespread in ponds of the Kafue River, Africa (Carey, 1967). Reiss (1977) showed two peak periods of abundance in the littoral benthos of the blackwater Tupe system in the Amazon. 2,559 animals/$m^2$ were collected on the rising flood, while 1,248 animals/$m^2$ were collected at low water; interim density was 623/$m^2$. Chironomids dominated the fauna at low water (but mites, corixid bugs and oligochaete worms were also present) while phantom midges (Chaoborinae) and ostracods replaced them as the water level rose. In whitewater floodplain pools, similar double peaks have been recorded, one before flooding and a second after high water. Benthos in the igapo forests is sparce (Sioli, 1975).

Fittkau *et al.* (1975) proposed three methods by which the benthos could survive between flood droughts:
(1) migration, principally seen in the large decapod crustaceans;
(2) dormancy, principally the method seen in the molluscs;
(3) recolonization by aerial adults, as seen in the insects.

Welcomme (1979) divides the fish species of tropical floodplains into two groups based on their behavioural responses to the fluctuating environment:

(A) Those fishes that avoid the fluctuations by migrating to and from the main river channel, e.g. species of Cyprinidae (minnows), Characoidei (characins), and some species of Siluridae (catfishes) and Mormyridae (Nile pikes).

(B) Those fishes that are able to survive the fluctuations, particularly that of dissolved oxygen, e.g. species of Ophiocephalidae (snakeheads), Anabantidae (climbing perches), Osteoglossidae, Polypteridae (bichirs), Protopteridae (lungfishes) and most Siluridae.

Although there are a large number of species in this latter category, very few of them are specifically adapted to survive complete loss of water from their habitat. If this occurs, then there is a huge fish kill. Only the lungfishes, some cyprinodonts and some mussels appear to be able to survive (see Chapter 5).

Fish production and biomass tend to vary according to the flood characteristics of specific areas. Very high floods cause the main channel water to spread out over the floodplain and this enables fishes to forage farther afield. Feeding rate is greatest at times of inundation and the fishes store what they assimilate as fat which lasts them through the lean dry period. Breeding, in most species, also coincides with the rise in water level and a smooth and gradual rise, with a high, sustained peak produces higher numbers of young-of-the-year than lesser floods. Thus in good years, young fishes emerge at a time when microorganisms, plankton, benthos and aquatic vegetation are abundant (Awachie, 1981). Floodriver fishes tend to have

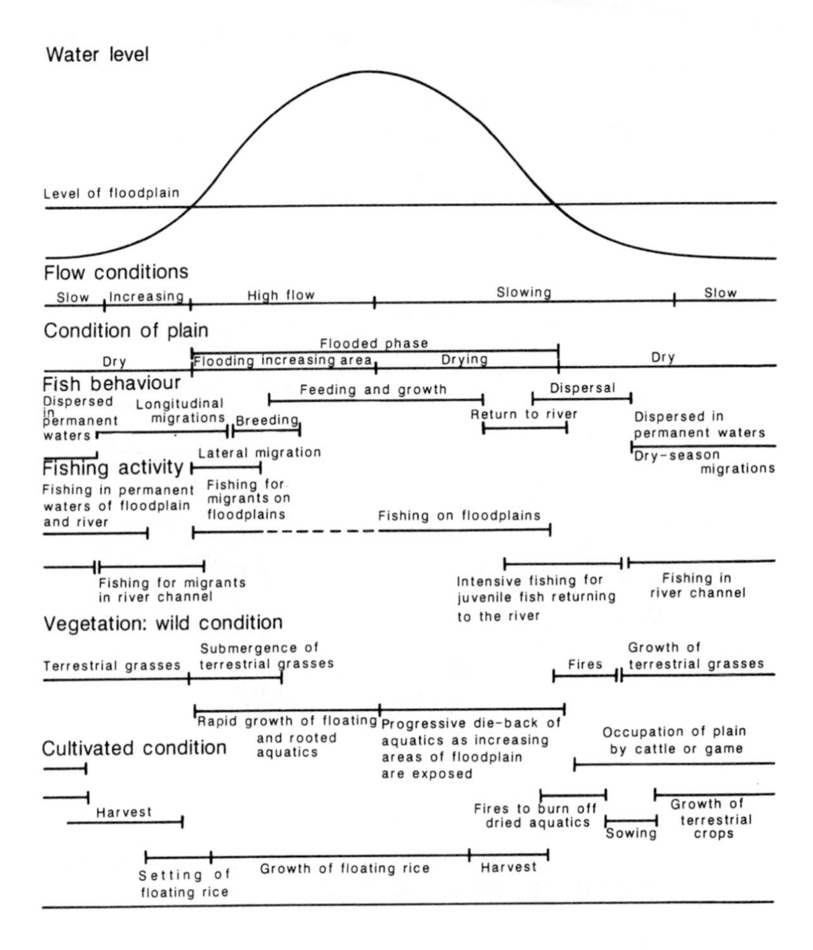


Figure 8.6: The main annual activities seen on natural and cultivated floodplains (redrawn from Welcomme, 1979).

short life cycles; many of them mature in one year and spawn at the next flood (Lowe-McConnell, 1977).

Many of these fish species are subject to heavy fishing by native peoples. However, this tends to be seasonal as, for example, during high water the fishes are spread out over a wide area of floodplain. Fishing effort tends to be concentrated at the time when the flood recedes and the water starts to drain back into the main channel; migrating species are caught in this way. Heavy fishing continues, during the dry season, in the basins of the floodplain, and also occurs at the beginning of the next flood cycle so as to again trap migrating species, this time on their way out of the main channel (Welcomme, 1979).

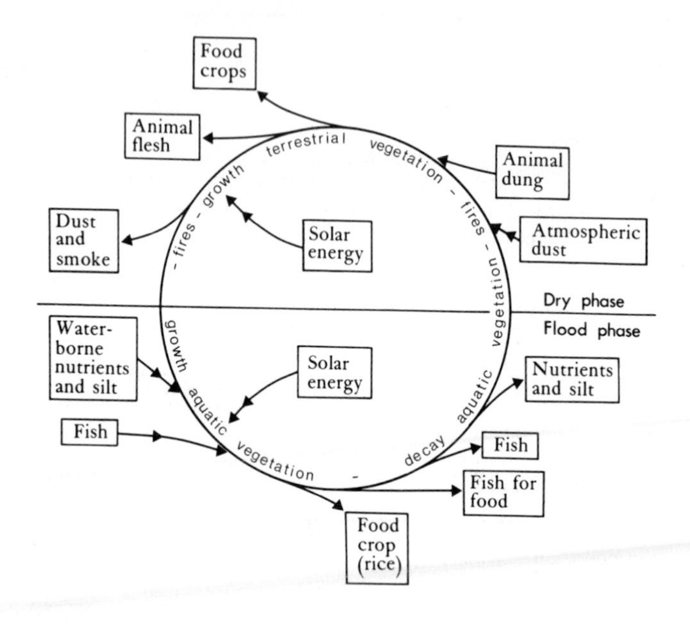


Figure 8.7: Energy and nutrient cycling in African floodplain systems (redrawn from Welcomme, 1979).

In many parts of the world, floodplains are under cultivation by man during the dry phase and so the system may parallel that of the Dombes ponds described earlier in this chapter. A summary of the main activities that occur on both natural and cultivated floodplains is given in Figure 8.6, and a scheme of nutrient and energy cycling in the system is given in Figure 8.7. Provided a natural community of plants and animals remains, the system appears to be able to function under both regimes. However, man's increasing appetite for land for development may seriously jeopardize its existence, particularly if it includes flood control, for cyclical flooding is the underlying factor upon which these complex ecosystems have developed.

## 8.3 Habitats for vectors of disease:

As previously mentioned, temporary waters have their deleterious side in that, particularly in the tropics, they may support populations of disease-carrying invertebrates.

*Schistosoma* is a genus of blood flukes that causes the condition known as schistosomiasis, or bilharzia, in man. Various species are common throughout many parts of Africa, South America and the Orient where their vector hosts are freshwater snails. Snail species suitable for the transmission of schistosomiasis (generally of the family Planorbidae) are common and pick up the ciliated larvae from a variety of shallow water bodies such as storage ponds, marshes, streams and banana drains (Sturrock, 1974). These miracidia secrete histolytic enzymes from special glands and this enables them to bore into the snails' tissue. Once inside the snail, each miracidium absorbs food from the host and after two weeks produces daughter sporocysts. After a further 2-4 weeks the cysts produce cercariae which then leave the snail and swim about in the water until they come into contact with the final host, often man. The usual method of entry into the body is by the animal forcing its way through the skin. The most important species are *Schistosoma mansoni, S. haematobium* (both occurring in the Middle East and much of southern and equatorial Africa; *S. mansoni* also occurs in parts of the West Indies and South America) and *S. japonicum* (common in the Far East). In *S. mansoni*, perhaps the most studied species, adults live in those portal system veins that drain the colon. The eggs are laid in small veins and develop into miracidia, some of which find their way into the intestine and thence into water when the host defecates near a pond or marsh. Symptoms of the disease include rashes, fever, inflammation, pain and damage to internal organs such as intestine, liver, spleen, lungs, bladder, spinal cord and brain.

In tropical Africa, *S. haematobium* is chiefly carried by snails of the genus *Bulinus* and is commonly endemic in areas of temporary standing waters which disappear in the dry season. It is clear that not only can the snails survive drought but that the schistosomes within them survive also and are infective when the ponds refill after rain (Beadle, 1981). *Schistosoma mansoni* is known to survive in aestivating *Biomphalaria* snails in Africa and Brazil (Barbosa and Barbosa, 1958; Jordan and Webbe, 1969). It seems likely that the larvae stop their development when the snails go into aestivation and, in this state, they can survive for up to seven months in snails not in water

(Richards, 1967).

Another serious condition caused by trematodes is fascioliasis. This is primarily a disease of the liver of cattle, sheep and other grazing animals but it may also affect man. Again, the intermediate hosts are snails, typically those living on the wet mud at the edges of small ponds, seeps and marshes (e.g. *Stagnicola bulimoides* and *Fosaria modicella*). They particularly thrive in temporary pools where they are capable of surviving the dry period by aestivation (Olsen, 1974). Eggs are deposited in the water in the faeces of grazing animals and only hatch after contact with the water. Again, it is the miracidium stage in the life cycle that seeks out a snail host. Development in the snail culminates in cercariae which emerge and encyst to form metacercariae. They are grazed off herbage by the final host. In *Fasciola hepatica*, the young fluke emerges from its cyst in the duodenum below the opening of the bile duct where the bile triggers emergence. The metacercariae initially feed on the mucosal lining and eventually migrate to the liver. The flukes mature and begin to lay eggs any time from 9 to 15 weeks later, depending on the host species. Symptoms of the disease include loss of weight and condition, swelling, breathing difficulties, toxemia and, often, death.

Another form of fascioliasis is caused by *Fasciolopsis buski*, a fluke common in eastern Asia and the southwest Pacific where it may affect as many as 10 million people causing intestinal and toxic symptoms. The life cycle is similar to that of *F. hepatica* but *F. buski* lives in the duodenum. After a sporocyst generation and two generations of rediae, cercariae are produced which encyst on the surfaces of aquatic vegetation such as water caltrop and water chestnut. When the seed pods or bulbs of these plants are harvested and eaten, they are invariably first peeled using the teeth. It is during this process that the metacercarial cysts are ingested (Jones, 1967).

*Paragonimus westermani*, the oriental lung fluke, infects man and a number of domestic and wild animals. It has two aquatic intermediate hosts. The snail (e.g. *Pomatiopsis lapidaria*) is penetrated by miracidia which swim about in marshy seepages and floodplain pools after having been deposited in faeces. Sporocysts develop in the snail and produce rediae which migrate into the lymph system of the liver. The rediae contain mature cercariae which emerge in the late afternoon - early evening to coincide with the peak activity time of their second intermediate host, either freshwater crayfishes or crabs. In these secondary hosts, the cercariae encyst around the heart where metacercariae are produced but not released until the crustacean is eaten by the definitive host which is, in many cases, man.

A number of nematode parasites are associated with temporary water bodies. *Dracunculus medinensis* is a parasite of the subcutaneous tissues of man and it occurs in many parts of the tropics, particularly in Africa and Asia where it is known as the guinea worm. It is typically a parasite of rural populations in arid regions where household water is obtained from step wells and small ponds (Olsen, 1974). The larval nematode lives in copepods such as *Cyclops leuckarti* and *C. hyalinus* together with species of *Eucyclops, Mesocyclops* and *Macrocyclops*. If a copepod becomes infected it tends to sink and, in wells, this stage seems to coincide with the period of drought - this may increase the probability of infection by concentrating the infective larvae in a relatively small volume of water (Croll, 1966). Infection of man occurs when these copepods containing fully-developed third stage larvae are swallowed in drinking water. It may be 10 to 14 months before mature females produce young.

Many tropical disease organisms are spread by mosquitoes that breed in temporary ponds and container habitats. *Wuchereria bancrofti* is a nematode worm that causes either mumu or elephantiasis, depending on the severity of the infection. Mumu is a painful swelling of the lymphatic ducts while elephantiasis is a widespread blockage of lymphatic channels from any distal portion of the body. *Wuchereria bancrofti* is found in Polynesia as well as in most of the tropical world and is transmitted to man by any of several species of mosquito but particularly those belonging to the genera *Culex, Aedes* and *Anopheles* (Jones, 1967). Microfilariae are ingested by female mosquitoes, as they take a blood-meal from an infected person, and penetrate the stomach lining to lodge in the insect's thoracic muscles. Here they transform into robust infective larvae which then migrate (some 8-14 days later) to the mosquito's proboscis in readiness for transfer back to a human host. In certain regions of the tropics, virtually all the inhabitants of a particular village or community may have larvae in their blood (Jones, 1967).

One of the world's most serious diseases is malaria. It is caused through an infection by the protozoan *Plasmodium* of which there are many species but *P. vivax* is perhaps the most widespread. *Plasmodium vivax* is a parasite of human liver and red blood cells, and is spread through the bite of many species of anopheline mosquitoes. It occurs in temperate as well as tropical climates. Other forms of malaria, in humans, are caused by *P. falciparum* (subtropics and tropics) and *P. malariae* (tropics, except South America) while other species in the genus may be specific to non-human primates and birds (Olsen, 1974).

The symptoms of malaria are well-known, chills, recurrent fever, weakness and emaciation. It affects hundreds of millions of people, directly, through debilitation but many more, indirectly, through losses in labour and efficiency. The disease has been known for at least 2,500 years and Hippocrates is credited with having made the connection between the occurrence of malaria and the proximity to swamps and stagnant waterbodies. Mosquitoes that carry malarial parasites include the genera *Culex, Aedes, Culiseta, Psorophora, Armigeres, Mansonia* and *Anopheles* but only the latter is responsible for spreading human malaria. However, there are about 40 species of *Anopheles* that can transmit *Plasmodium vivax* alone and these include species found in Europe and North America.

Survival of the mosquitoes when their larval habitat dries up is often dependent on gravid females. Females of *Anopheles gambiae*, the most important of the African vectors, frequently aestivate in a dormant state in crevices (e.g. in walls, in dry wells and in rodent burrows) during the long dry season. Omer and Cloudsley-Thompson (1968) showed that, in the Sudan, 77.1% of female *A. gambiae* in this condition were engorged with blood (90.6% of it human), and that their ovaries developed very slowly during the drought so that when the rains came the females were ready to oviposit.

Another vector species of mosquito, *Aedes aegypti*, survives drought as resistant eggs which are laid on the damp margins of waterbodies as they are drying. *Aedes aegypti* transmits the virus known as yellow fever which was, in the past, a major disease of man. It was particularly common in port cities and was spread by trading sailing ships, sick seamen and the mosquito which often bred in water barrels on board the ships (Jones, 1967).

*Aedes aegypti* is responsible for transmitting yet another virus, dengue, or "break-bone fever". The pattern of infection is very similar to that of yellow fever with old trading ships acting as breeding places for the mosquito. As a new port was reached, the virus got ashore either in the blood of an infected seaman or in the salivary glands of an infected mosquito. It then became established through local susceptible hosts and local *Aedes* and an epidemic would typically follow. As with yellow fever, dengue is less of a problem now than in the past. However, reservoirs of both these diseases exist in populations of monkeys in many tropical jungles. Related to dengue are the various forms of viral encephalitis which break out periodically in the United States, Canada, Australia, Japan and other countries. These affect the human central nervous system and are potentially very dangerous. Many birds carry enough of the virus in their blood to act as reservoirs for the disease and transmission to man is effected by quite a number of mosquito species, chiefly those belonging to the genera *Culex, Anopheles, Aedes* and *Psorophora* (Jones, 1967).

In addition to breeding in larger temporary waters, many vector species lay their eggs in phytotelmata - those small, intermittent waterbodies associated with plants (see Chapter 7). This is especially so for the mosquitoes.

In tropical Africa, sylvatic yellow fever is maintained and circulated nowadays, chiefly amongst animal populations but is sporadically passed on to man. As we have seen, the primary vector is *Aedes aegypti* but it can be transmitted also by five other species, and a further eleven species, while not yet proven vectors, could be vectors. Of this total of 17 species, 12 belong to the genera *Aedes* or *Eretmapodites* and all, except for those of the *Aedes dentatus* group, either normally or occasionally have larvae capable of being reared in common types of phytotelmata, particularly treeholes and leaf axils. Availability of water-filled leaf axils has been shown to be influenced by agriculture, as crops such as banana, plantain, pineapple and cocoyam form particularly good water reservoirs and are the principle habitats of *A. simpsoni* (Pajot, 1983). Populations of *Anopheles neivai* in Columbia, *A. bellator* and *A. homunculus* in Brazil and Trinidad, and *A. cruzii* in Brazil are proven carriers of human malarias. *Anopheles bellator* is also known to carry the nematode *Wuchereria brancrofti.* All these mosquitoes breed in phytotelmata (Frank, 1983).

Other important human diseases spread by insects that are associated with temporary waters or moist habitats during some stage in their life cycles include leishmaniasis (Kala-azar of most tropical and subtropical regions) transmitted by sandflies (Psychodidae); other forms of filariasis (in African rainforests) transmitted by "no-see-ums" (Ceratopogonidae); Loa loa (in Africa) transmitted by tabanid flies (Tabanidae); pinkeye, conjunctivitis and trachoma (in the New World) transmitted by eye flies (Chloropidae); and trypanosomiasis (African sleeping sickness) transmitted by tsetse flies (Muscidae) (Jones, 1967; Lainson and Shaw, 1971).

Control of many of these diseases usually involves identification of the specific vector and erradication of either it or its habitat. For those vectors such as freshwater snails and mosquitoes this may involve application of molluscicides and insecticides, drainage of wet areas of land, introduction of predators (e.g. dytiscid beetles or fishes), or controlled raising or lowering of water levels at crucial times in the life cycle. In southern Brazil, deforestation around towns was effective control of *Anopheles* because it removed the epiphytic bromeliads that were providing habitats for the larvae. Although flight ranges of *A. bellator* and *A. cruzii* are known to be in excess of 2km, few adults crossed the 1km deforested zone. In Trinidad, the incidence of malaria decreased after the bromeliads of shade trees around cacao plantations were destroyed by herbicides (Smith, 1953). However, some of the other insects that live in bromeliads may be important pollinators of both crops and natural vegetation so that widespread destruction of phytotelmata may be a disadvantage (Frank, 1983). No single method may be totally effective, particularly as many species of mosquito become resistant to chemicals quite quickly, and so very often a combination or succession of techniques, intelligently applied, is required.

# 9 TEMPORARY WATERS AS STUDY HABITATS

The ubiquitous nature and small size of temporary ponds and streams, combined with high degree of physiological tolerance of many of their inhabitants, make them ideal subjects for practical classroom study. Not only can they be used to illustrate fundamental features of aquatic systems but they are perhaps the best environments in which to study species interactions and community succession. In addition, they lend themselves perfectly to experimental investigation of colonization mechanisms and physiological tolerance limits.

There follows a list of suggested exercises which may be undertaken by students. Space does not permit detailed outline of experimental procedures but guidelines are given where appropriate. To a large extent, the scale of each experiment will depend on resources at hand and the time available in the classroom together with the accessibility of temporary water habitats. Some of the exercises are relatively short-term and can be completed through intermittent attention by a single student over just a few laboratory sessions. Others are longer-term and more labour-intensive but lend themselves to cooperative work. The apparatus required varies between exercises but is commonplace in most laboratories; any special requirements are indicated. For further reading on some specific topics, one or two pertinent references are appended. Where identification of organisms is required, the reader is referred to keys appropriate to that country - in Britain, for example, there is a series of guides to the identification of British freshwater organisms published by the Freshwater Biological Association, while in North America, there are texts such as "Freshwater Biology" (Edmondson, W.T., ed. 1959, Wiley & Sons, N.Y., 1248 pp). Interpretation of the significance of results obtained from these exercises is left largely to the reader, however prior reading of the other chapters in this book will help.

## 9.1 Investigations of the structure and function of aquatic populations and communities:

*(a) Determination of a population growth curve for a temporary pond species and response to environmental variation:*

Growth curves are often used as indices of population dynamics. A population of a species growing in a predator-free environment where resources are infinite would show exponential growth. In nature this never happens as, inevitably, one or more resources become depleted and growth is limited. Typically the population levels off at a plateau although, in practice, slight, continual adjustments between the animals and the resources result in minor oscillations about this level (Figure 9.1). Temporary pond species, particularly zooplankton, are well suited to illustrating this phenomenon.

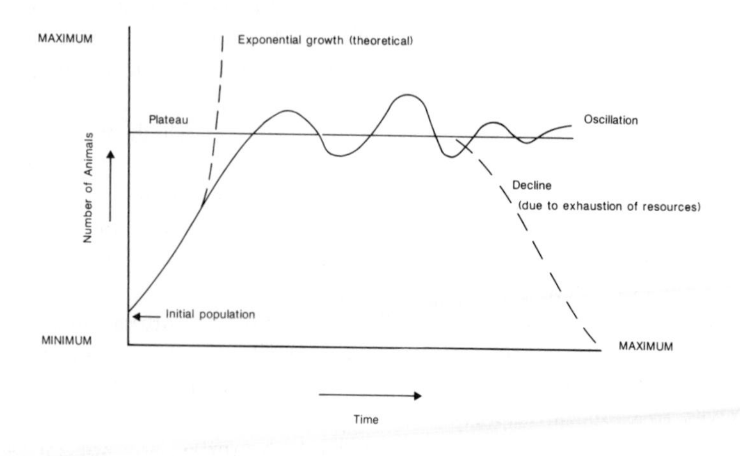


Figure 9.1: Generalized population growth curve for a species.

*Procedure* - Collect, from a temporary pond, several hundred individuals of a single species of cladoceran such as the cosmopolitan *Daphnia pulex* or species from other widespread genera (*Ceriodaphnia, Simocephalus, Chydorus*). Place 50 specimens in a one litre container of synthetic pond water (prepared by dissolving 80mg $CaCl_2$, 50mg $KHCO_3$, 50mg $NaNO_3$, 50mg $K_2PO_4$ and 24mg $MgSO_4$ in 1 litre of distilled water) to which has been added 100mg of dried yeast as food. The food must be kept in suspension, and the water oxygenated, by gently bubbling air into the water. To obtain a rapid response, the animals should all be ovigerous females. Assess the population density every four days by, after mixing, removing a 100ml subsample of the water with a wide-mouthed pipette and counting the animals in it (these should afterwards be returned to the population). Make a graph of the number of animals versus time. Repeat the experiment periodically changing the water and re-adding the same amount of food. How does the resulting graph differ from the first? Compare the slopes, plateaux and shapes of these two graphs with others obtained from populations subjected to environmental change, for example increase or decrease in water temperature, pH or food level. (*Reference* - Odum, E.P. 1971. Fundamentals of Ecology, W.B. Saunders, London, 546 pp).

*(b) Interspecific competition in cladocerans:*

Ecological principles dictate that only one species can occur in a particular niche at any one time. If, in a new environment, two species attempt to occupy the same niche then competition may arise and eventually one species may be eliminated. It is possible to demonstrate this experimentally although it is easier to see the end product than the actual competition process. *Note*: This exercise illustrates classical Gausian competition (Gause, 1934) in which, in simplified environments such as are produced in lab. experiments, one species usually ousts the other. In natural environments, however, it is now thought that many factors, both physical and biological, may act to keep the population levels of two species below a level at which they would begin to compete for a common, finite resource (e.g., food). For coverage of recent concepts and controversial issues on competition the reader is referred to Strong et al. (1984).

*Procedure* (from Wratten, S.D. and Fry, G.L.A. 1980. Field and Laboratory Exercises in Ecology. Edward Arnold, 227 pp). - Collect sufficient numbers of ovigerous females of two closely related species, e.g. from the genus *Daphnia*. Using artificial pond water (see recipe in previous exercise), set up the experimental combinations shown in Table 9.1 in 200ml containers. Suspend the yeast by stirring, add the appropriate number of animals to each container using a pipette and label each container. Remove all the animals from each container every four days and count them before replacing them in a fresh container with the appropriate water and food levels. Enumeration should continue until either a steady state is reached or a population becomes extinct. Plot population levels against time for each combination and assess the impact of food level, space and interspecific competition on population development.

Table 9.1: Combinations of density, species, habitat size and food level needed to experimentally examine interspecific competition in cladocerans.

| Experiment | Species | Number of animals added | Volume of pond water (ml) | Amount of food added (as mg of dried yeast) | Suggested number of replicates |
|---|---|---|---|---|---|
| 1 | A | 5 | 100 | 5 | 10 |
| 2 | A | 5 | 100 | 10 | 10 |
| 3 | A | 5 | 100 | 20 | 10 |
| 4 | A | 5 | 50 | 10 | 10 |
| 5 | B | 5 | 100 | 10 | 10 |
| 6 | A + B | 5 + 5 | 100 | 10 | 10 |

*(c) In situ measurement of the population dynamics of a temporary pond cladoceran with identification of the different morphs:*

As we saw in Chapter 5, temporary water cladoceran females rapidly build up their population numbers at certain times of the year through parthenogenesis. At other times of the year, generally before the dry period, "ephippial", "resting" or "winter" eggs appear which are the result of sexual reproduction, necessitating the presence of males in the population.

*Procedure* - Take standardized weekly samples of zooplankton from a temporary pond by sweeping a fine-mesh net (approximately 100 micron mesh openings) through a fixed arc in midwater. Preserve the samples in 10% formalin or 70% ethyl alcohol and label with the date and water temperature. In the laboratory, pick out all the individuals of the required species under a microscope and count the numbers in each of the following categories: parthenogenetic female, ephippial female, male, juvenile (see Figure 9.2). To reduce the time spent sorting through these samples, particularly if there are large numbers present, each sample may be subsampled following the procedure outlined in Elliott (1977, page 135. Some methods for the statistical analysis of samples of benthic invertebrates. Freshwater Biological Association of the U.K. Scientific Publication 25, 160 pp). Plot the total number of animals in each sample against time, water temperature and amount of water in the pond (or pond diameter). Note the times of first occurrence of the different morphs along these graphs and their subsequent representation (% occurrence) in the sample.

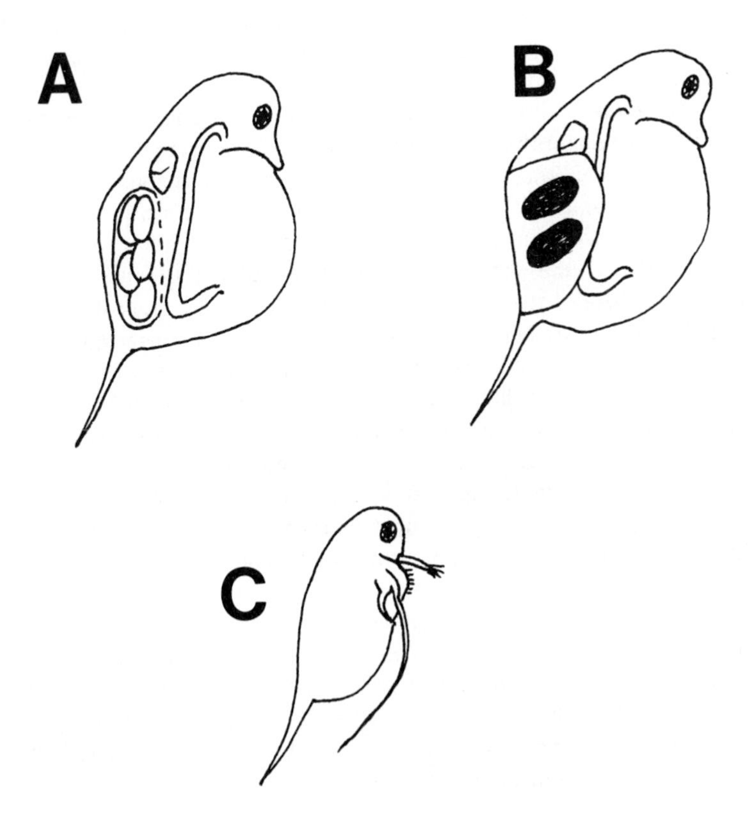


Figure 9.2: Different sexual morphs of *Daphnia:*
(A) Parthenogenetic female showing eggs in brood-sac;
(B) Ephippial female showing two eggs contained in ephippium;
(C) Male, note smaller size, larger first antennae and the stout hook on the first limb (redrawn from Scourfield and Harding, 1966).

*(d) Comparison of the changes in community composition along a hydroseral gradient:*

From Chapter 4 it is evident that distinct sets of physical and chemical conditions are characteristic of habitats along the seral progression from temporary pond to dry land. In certain localities it will be possible to find examples of all these habitats in a relatively small area thus allowing comparison of the species aggregations living in each.

*Procedure* - Although potentially a large scale exercise extending over many months, this study can be abbreviated by sampling only three or four stages along the sere only once in the year - preferably the wettest part. Stages chosen might, for example, be: a pond holding water for eight months of the year, a pond holding water for four months of the year, a marsh or fen, and a bog with shallow pools. The flora of these habitats can be collected by hand while the macrofauna is best sampled by handnet (100 micron mesh openings) from both the water column and bottom substrate. Samples should represent all the microhabitats evident and sampling effort should be constant between habitats. In the laboratory, pick the organism from the debris under a dissecting microscope and record the species (or visually different types) together with a semiquantitative estimate of abundance (see Figure 4.7). Faunal lists can then be compared and changes in community composition noted. Trends detected, for example, might include an increase in the abundance and diversity of larval craneflies (Diptera; Tipulidae) or enchytraeid worms as the amount of water diminishes or, conversely, a decrease in the diversity of Coleoptera or chironomid midges. (*Reference* - Caspers, H. and Heckman, C.W. 1981. Ecology of orchard drainage ditches along the freshwater section of the Elbe Estuary. Archives für Hydrobiologie/Suppl. 43 (4): 347-486).

*(e) Morphometric comparisons of identical species from temporary and permanent waters:*

There is some evidence to indicate that marked differences in size (and in the case of copepod crustaceans, the number of eggs/female) occur between populations of the same species in the two habitats.

*Procedure* - Locate two habitats, one temporary the other permanent, that have a number of species in common. Sample both populations at the same time of year and preserve them in 10% formalin or ethanol. It does not matter which species are chosen for analysis although those with a hard exoskeleton (i.e. arthropods) are preferable to soft-bodied forms. Measure at least 100 individuals chosen at random, from each sample. The parts of the animals to be measured will vary depending on the group chosen. For example, for copepods measure the length of the cephalothorax along the midline of the body, for cladocerans measure the height or width of a single valve, and for insect larvae measure the width of the head at its widest point. These parameters are preferred to overall animal length as this can alter during preservation due to contraction of the softer regions of the body. If eggs are present in female crustaceans they should be counted. Males and females should be analyzed separately if they are distinguishable in order to remove differences due to sexual dimorphism. Construct cephalothorax length-frequency or headwidth-frequency histograms as appropriate for each species in each habitat. If differences are not immediately obvious, the data may be treated statistically so as to test for significant differences. To do this, use either the non-parametric Mann-Whitney U-test to compare the data sets (if data points are not normally distributed) or Student's t-test applied to the respective mean values (if the distribution is normal) (See Elliott 1977, pages 95 and 112). Interpret the results obtained in terms of the differences between the two habitats with respect to competition, predation, food supply, etc. (*Reference* - Cole, G.A. 1966. Contrasts among calanoid copepods from permanent and temporary ponds in Arizona. American Midland Naturalist 76 (2): 351-368).

*(f) Determination of the basic elements of the food web of a temporary pond or stream:*

Energy is transferred through a community by one organism eating another. In the very simplest situation this may resemble a series of links in a food chain. In most communities, however, interactions - such as one prey species being eaten by two species of predators - cause branches in the system likening it to a web. Some food webs are very intricate and therefore time-consuming to analyze. Nevertheless it is possible to determine just the major links in a web by examining only the most abundant species and some temporary waters are relatively simple systems so they are especially appropriate for this kind of study.

*Procedure* - Take one large net sample (100 micron mesh) from a pond or stream, making sure that the major microhabitats (e.g. bottom debris and water column) are represented. Preserve the sample immediately to arrest digestion in the animals. In the laboratory, draw random subsamples for examination under a dissecting microscope. Pick out only those species that are well represented numerically or in terms of size. Preserve these animals in 70% ethanol. The

Microdissection in a cavity slide

↓

Slit body wall and remove entire gut

↓

In distilled water, slit gut wall and remove gut contents, discard gut wall

↓

Break up gut contents with forceps and wash into 50 ml beaker

↓

Mix thoroughly using forceps or magnetic stirrer

↓

Wash into Millipore funnel and filter column fitted with a 0.45 μm gridded filter

↓

Filter at low (<5psi) vacuum, terminate filtration while filter is still damp

↓

Using forceps, place filter on 1–3 drops of light immersion oil, add 1–3 drops to filter surface; keep covered until filter becomes transparent (approximately 24 hours)

↓

Blot excess oil from beneath filter; add 1–2 drops of clear mounting medium, then a coverslip

↓

Under a highpower microscope, note the major food items (eg. detritus, algae, whole small animals, fragments of larger animals). Identify them as far as possible and make an estimate of their relative abundance

Figure 9.3: Generalized procedure for analyzing the gut contents of aquatic invertebrates (abbreviated from Cummins, 1973).

easiest way of piecing the food web together is through analysis of the gut contents of the major species. A generalized procedure for this is shown in Figure 9.3. Analyze the gut contents of at least 10 individuals of each of the major species. In some instances, due to the nature of an animal's feeding technique (e.g. sucking) or subsequent digestion process (e.g. grinding in a gastric mill), identification of food items may be difficult. In these cases, in laboratories suitably equipped, other techniques may be tried (e.g. Davies, R.W., Wrona, F.J. and Linton, L. 1979. A serological study of prey selection by *Helobdella stagnalis* (Hirudinoidea). Journal of Animal Ecology 48: 181-194). Upon completion of the gut analyses, the most likely food web for the pond or stream community can be constructed. This, however, will indicate only the most obvious links between the major species at just one time of year. With more effort, the study could be extended to encompass seasonal differences which, in the case of temporary waters with their rapid seasonal succession of species may be considerable. (*References* - Jones, J.R.E. 1949. A further ecological study of calcareous streams in the 'Black Mountains' district of South Wales. Journal of Animal Ecology 18: 142-159 - for a simple example of an aquatic food web).

*(g) Predator-prey interactions:*

Population fluctuations of a species may be the result of both internal and external factors and one important example of the latter is predation. In temporary freshwaters there are many obvious and varied examples of predator-prey relationships which lend themselves to study. For example, copepods such as *Diaptomus* and *Acanthocyclops* prey on cladocerans. Periodic sampling of both predator and prey populations (as outlined in Exercise C) over the aquatic phase of a pond may produce data which when plotted as a graph produce a classic picture of cyclic oscillation in population densities with the peaks and troughs in predator numbers following (with a slight lag) those of the prey. Such a pattern will be most obvious where the relationship consists of just one prey and one predator species. Population decline in some prey species may coincide with the influx of predatory migrants, for example adult water beetles or hemipterans that feed on mosquito larvae or fairy shrimps. Care should be taken, however, in attributing all population fluctuations seen in prey species to predators.

Lightheartedly, predation may be seen as a very personal experience and, therefore, some account should be made of it at the level of the individual. Direct observation of predator attack behaviour and subsequent prey evasion tactics can be made quite easily in the laboratory through a horizontally-mounted dissecting microscope with or without the aid of a film camera. (For procedure and examples the reader is referred to Kerfoot, W.C. 1978. Combat between predatory copepods and their prey: *Cyclops*, *Epischura* and *Bosmina*. Limnology and Oceanography 23: 1089-1102).

## 9.2 Investigations of colonization patterns, dynamics and mechanics of temporary water species:

*(a) Determination of seasonal and/or diel periodicity of colonization of temporary waters by aquatic Coleoptera and Hemiptera:*

Some species of beetles and true bugs commonly occur opportunistically in temporary waters (see Chapter 6). They fly in from permanent ponds to feed and leave again when conditions become unfavourable.

*Procedure* - Place one large or several small artificial ponds (e.g. a plastic paddling pool or plastic bowls, respectively) in an open area in early spring and fill with water. For seasonal analysis, collect all the insects that have colonized the "ponds" two or three times each week using a fine-meshed net. For diel analysis, collect all the insects at two or four hour intervals throughout the day and night on several occasions throughout the year. Preserve all specimens in 70% ethanol to await identification and counting. Measure water and air temperatures and note the weather conditions when sampling. For each species draw a histogram relating numbers to time (if there are insufficient numbers, the data can be pooled). Is there a relationship between flight periodicity and environmental conditions? (*Reference* - Fernando, C.H. and Galbraith, D. 1973. Seasonality and dynamics of aquatic insects colonizing small habitats. Internationale Vereinigung für Theoretische und Angewandte Limnologie Verhandlungen 18: 1564-1575).

*(b) Analysis of the environmental cues that attract colonizing species to temporary waters:*

Actively colonizing species are attracted to temporary waters for food, oviposition or both. Species respond to different cues as they locate the precise position of the pond and review its characteristics. Female mosquitoes, for example, respond to visual, tactile and chemical stimuli when choosing the optimum type of habitat for successful development of their larvae. Determination of some of these cues is possible by varying the environmental characteristics of the "ponds" used in the previous exercise.

*Procedure* - Using a large number of small bowls (50cm diameter) filled with water and placed in the open, vary the following features, either singly or in combination: *1)* colour of bowl - try a light vs a dark colour. *2)* pH of the water - by adding acid or alkali, try a pH of 6.0 vs one of 8.0. *3)* Chemical composition of the water - make an infusion of grass and add it to one set of bowls. *4)* Characteristics of the "pond" margins - place a ring of upright cut vegetation from a local pond around the outside of the bowl. Ensure that this material is free of animals beforehand.

Maintain a sufficient number of control "ponds" so as to allow for meaningful comparison of the results. The "ponds" should be left *in situ* for at least two weeks during which they should be regularly topped up with water. At the end of this period, empty each bowl through a 100 micron mesh net and remove any animals to 70% ethanol. Count and identify both adults and larvae. If statistical treatment of the data is desired, then a multiple range test (such as Newman-Keuls or Duncan's) is most appropriate. This exercise can be adapted for the study of temporary stream species by using sections of eavestrough through which water is recirculated (see Williams, D.D. and Hynes, H.B.N. 1976. Stream habitat selection by aerially colonizing invertebrates. Canadian Journal of Zoology 54: 685-693). (*Reference* - Gubler, D.J. 1971. Studies on the comparative oviposition behaviour of *Aedes (Stegomyia) albopictus* and *A. (S.) polynesiensis* Marks. Journal of Medical Entomology 8(6): 675-682).

*(c) Determination of the cues that cause adult animals to leave temporary waters:*

After feeding or reproducing in a temporary waterbody, some species fly off to another location. This exercise examines, in the laboratory, which stimuli initiate dispersal.

*Procedure* - Collect a selection of adult Coleoptera and Hemiptera from a pond. Particularly suitable are members of the families Dytiscidae, Hydrophilidae, Gyrinidae and Haliplidae (Coleoptera) and Corixidae, Notonectidae and Nepidae (Hemiptera). Place monospecific batches of five individuals in small (25cm diameter) bowls of pond water and subject the water to the following changes: *1)* gradual warming from pond temperature to $40^{o}C$. *2)* Gradual increase in conductivity - by adding small amounts of a strong solution of the salts outlined in Exercise 9.1.a. *3)* Gradual increase in pH - by adding a suitable alkali. Record the activity level of the animals along these gradients and note when they come to the water surface and take flight. (*Reference* - Landin, J. and Stark, E. 1973. On flight thresholds for temperature and wind velocity, 24-hour flight periodicity and migration of the water beetle *Helophorus brevipalpis* Bedel (Coleoptera: Hydrophilidae). Zoon, Supplement 1: 105-114).

*(d) Analysis of passive dispersal mechanisms of micro-organisms from temporary waters:*

Not all temporary water inhabitants are capable of dispersing to other habitats under their own steam. Many are passively transported by wind or larger animals but in order for these techniques to be successful the species have to have eggs or cysts which can withstand drying and remain viable for quite some time.

*Procedure* - (based on Maguire, B. 1963. The passive dispersal of small aquatic organisms and their colonization of isolated bodies of water. Ecological Monographs 33: 161-185). Select a pond greater than 10m in diameter. Clear bankside vegetation to a height of 10cm along a path approximately 5m wide and 50m long at right angles to the pond on the downwind side. Establish stations at 1, 2, 5, 10, 20 and 50m along the middle of the path. At each station, erect a steel post 1.5m high. Attach glass jars (approximately 10cm wide by 8cm deep) to each post at heights of 0.5, 1.0 and 1.5m; each jar having been previously sterilized and filled with sterile, artificial pond water. In areas of intense sunlight it may be necessary to partially shade the sides of the jars with aluminum foil so as to prevent the temperature of the water from becoming unduly high. Water lost through evaporation should be replaced regularly with double-distilled water. Avoiding contamination, take one 25ml sample from the stirred contents of each jar at two week intervals and preserve them. Filter these samples onto 0.45 micron gridded filters and clear and mount them as outlined in Exercise 9.1.f. Contrast the numbers and types of organisms found (using a highpower microscope) with respect to distance from the pond and height shown above the ground. If time and manpower permit, make a brief survey of the microorganisms in the pond for comparison.

## 9.3 Investigations of the physiological adaptations of temporary water faunas:

*(a) Comparison of physiological tolerances of closely related species from permanent and temporary habitats:*

Chapter 5 contains many examples of the extreme physiological tolerance of temporary water species. Some of these may be demonstrated in the laboratory using relatively unsophisticated equipment.

*Procedure* - Comparisons may be made on the basis of some or all of the following: *(1) water temperature* - particularly upper and lower lethal limits. For ULT, place individuals of a temporary pond species and a closely related permanent pond species, separately, into pond water in a series of large test-tubes (say 30 specimens of each). Place the tubes in a water bath and, starting at pond temperature, raise the temperature of the bath 1$^{o}$C every five minutes. Each time the temperature is raised, check the condition of each specimen and note when activity levels change and, ultimately, when death occurs. Plot the number of each species surviving against temperature and note the $LD_{50}$ - that is the temperature at which 50% of the specimens are dead. The ULT of some species can be extended if the animals are allowed sufficient time to acclimate to different temperature levels; however, this is unlikely under the suggested rate of temperature increase. Nevertheless, the experiment could be expanded to include a comparison of the acclimation capacities of the two species. For testing LLT, specimens can be subjected to a decrease in temperature to 0$^{o}$C or even freezing in ice for varying periods (checking for survivors upon thawing). *(2) pH levels* - add acid or alkali to vary the pH of the pond water and determine the $LD_{50}$. *(3) resistance to desiccation* - remove specimens of both species to filter paper and blot dry. Place them on fresh filter paper in an incubator held at 15$^{o}$C which contains several open dishes of water so as to produce a relative humidity of between 80 and 90%. Remove five specimens of each species at one hour intervals up to 24 hours. Immerse them in 15$^{o}$C pond water and check for signs of life. *(4) minimum dissolved oxygen levels* - this exercise requires the use of a laboratory oxygen analyzer equipped with a respiratory chamber of approximately 5ml capacity. Place a single large animal or a few small animals in stirred, oxygen-saturated water, at pond temperature, in the respiratory chamber. Continually record the decrease in oxygen in the chamber until the meter registers zero or a steady state is reached. Record the behaviour of the animals along this gradient and also their $LD_{50}$. If any remain alive at zero oxygen, keep them in deoxygenated water for varying periods in order to compare survival times. Dry the animals to constant weight in an oven and calculate oxygen uptake values in terms of mg/g dry wt/hr. Plot uptake values against decrease in oxygen in the chamber. Examine these graphs for any indication of a change from the animals' oxygen consumption being independent of oxygen concentration to being dependent on oxygen concentration in the water. (*Reference* - Moore, W.G. and Burn, A. 1968. Lethal oxygen thresholds for certain temporary pond invertebrates and their applicability to field situations. Ecology 49: 349-351. Harris, R.E. and Charleston, W.A., 1977. The response of the freshwater gastropods *Lymnaea tomentoso* and *L. columella* to desiccation. Journal of Zoology. London 183: 41-46. Berg, K., Jonasson, M. and Ockelmann, K.W. 1962. The respiration of some animals from the profundal zone of a lake. Hydrobiologia 19: 1-39).

*(b) Determination of the length of diapause and the factors that terminate it:*

*Procedure* - Collect some mud from the dry bed of a temporary pond or stream. Sort through it under a microscope and extract any eggs, larvae, pupae, adults, cocoons, cysts, etc. seen. Group these stages into similar types for testing. The stage in the life cycle in which diapause is exhibited varies between species. For example, in chironomid midges and stonefly nymphs it may be the larva, while in mosquitoes and fairy shrimps it is the egg. Length of diapause can be examined by simply keeping the animals under the environmental conditions that maintain diapause for increasing lengths of time; periodically giving subsamples the conditions necessary to break diapause and comparing viability with time. Environmental factors terminating diapause may vary considerably in complexity between species but, obviously, they are those that occur naturally at the appropriate times in the pond cycle. Eggs of the fairy shrimp *Chirocephalopsis bundyi*, for example, are triggered by low oxygen levels in the water as the pond fills and a similar stimulus is responsible for hatching of the eggs of the British woodland mosquito *Aedes punctor*. Parameters that might be manipulated in an attempt to break the diapause of other species are: intermittent removal of water, increase in carbon dioxide levels, temperature, photoperiod and water chemistry. (*Reference* - Fallis, S.P. and Snow, K.R. 1983. The hatching stimulus for eggs of *Aedes punctor* (Diptera: Culicidae). Ecological Entomology 8: 23-28).

*(c) Patterns in egg hatching of temporary water species:*

In unstable habitats where environmental conditions change unpredictably, hatching of all the eggs laid by an individual female at the same time may be disastrous - it is the epitome of "putting all one's eggs in the same basket". Many temporary water species therefore show staggered egg hatching, thus ensuring that at least some larvae will survive.

*Procedure* - (based on Woodward, D.B., Chapman, H.C. and Petersen, J.J. 1968. Laboratory studies on the seasonal hatchability of egg batches of *Aedes sollicitans, A. taeniorhynchus* and *Psorophora confinnis.* Mosquito News 28(2): 143-146). Collect newly emerged female mosquitoes from a temporary pond in the spring, summer or autumn using a net and pooter tube (or other aspirator). In the laboratory, provide these females with a blood meal in the form of a rabbit or guinea pig. Isolate each female in a 50ml glass vial provided with a moist cottonwool plug and a raisin. Each female may lay several batches of eggs and should be re-fed and removed to a fresh vial after each oviposition. Record the numbers of eggs/batch and batch sequences. Take each batch and provide it with hatching stimuli appropriate for the species (e.g. cover with water of low oxygen content) and record the number of eggs that hatch. Remove the remaining eggs to dry petridishes for several days and then re-stimulate. Continue this process until no further eggs hatch. What percentage of the eggs in a single batch hatch upon exposure to the initial and subsequent stimuli? Is there any difference in hatching pattern between the various batches laid by the same female or at different times of the year? This exercise could be tried with eggs of the fairy shrimp.

*(d) Plasticity of body form of Artemia in response to changes in environmental conditions:*

*Artemia salina*, the brine shrimp, is a cosmopolitan species in saline inland waters, many of which periodically dry up. It responds, in external appearance (for example the number and length of limbs), to changes in environmental conditions, particularly salinity and alkalinity of the water. These changes are thought to be adaptations to buoyancy control. Successive generations of individuals reared in water of low salinity tend to show an increase in the number and length of limbs (surface area). Conversely, in highly saline water, total body surface area may be reduced as the medium provides greater buoyancy.

*Procedure* - Obtain viable eggs of *Artemia* from a biological supply house and hatch them by immersing them in water of the recommended salinity. Split the larvae up into large containers (10 litres) of differing salinity and provide food in the form of a suspension of yeast (100mg/l). Rear several generations in each container, periodically removing subsamples for morphometric measurements. Calculate means of length, width and surface area of a variety of body parts together with the mean number of limbs/animal for each salinity. Test for significant differences using an appropriate statistical procedure.

*(e) Evaluation of the relative importance of different oversummering methods in the survival of members of a temporary water community:*

In Chapter 5, a figure summarized the various methods by which species survived the summer dry period. The purpose of this largely field-oriented exercise is to quantify the relative importance of these methods for a given habitat.

*Procedure* - Systematically examine all the microhabitats in and around a dry stream or pond bed for active, torpid or diapausing individuals belonging to the community. Large animals may be immediately obvious while smaller ones will have to be searched for in debris or samples of bottom mud under a microscope. Sort the animals into different species where practicable. In the case of eggs, identification may be impossible. Further, some species (particularly of Coleoptera and Hemiptera) may have temporarily left the habitat but may be accounted for by reference to a list of species taken during the aquatic phase. Draw up a list of the various survival methods and calculate the percentage contributions of each to the survival of the community as a whole.

The above suggestions for experimental and descriptive study of temporary waters represent just a few of those that can be attempted by students. After reading the rest of this book, it should not be difficult for either students or instructors to see other possibilities for research projects, especially those associated with the particular types of temporary waters close to their home base.

# References

ALDERDICE, D. F. 1972. Responses of marine poikilotherms to environmental factors acting in concert. pp. 1659-1722. *In* : Marine ecology (O. Kinne, ed.), Wiley-Interscience, London.

AMEEN, M. U. & T. M. IVERSEN. 1978. Food of *Aedes* larvae (Diptera: Culicidae) in a temporary forest pool. Arch. Hydrobiol. 83:552-564.

ARTS, M. T., E. J. MALY & M. PASITSCHNIAK. 1981. The influence of *Acilius* (Dytiscidae) predation on *Daphnia* in a small pond. Limnol. Oceanogr. 26: 1172-1175.

ASAHINA, E. 1966. Freezing and frost resistance in insects. pp. 451-486. *In* : Cryobiology (H. F. Meryman, ed.), Academic Press, London.

ASHMOLE, N. P., J. M. NELSON, M. R. SHAW & A. GARSIDE. 1983. Insects and spiders on snowfields in the Cairngorms, Scotland. J. Nat. Hist. 17: 599-613.

AWACHIE, J. B. E. 1981. Running water ecology in Africa. pp. 339-366. *In* : Perspectives in running water ecology (M. A. Lock and D. D. Williams, eds.), Plenum Press, N. Y.

AYENI, J. S. O. 1977. Waterholes in Tsavo National Park, Kenya. J. appl. Ecol. 14: 369-378.

BAGNOLD, R. A. 1954. The physical aspects of dry deserts. pp. 7-12. *In* : Biology of deserts (J. C. Cloudsley-Thompson, ed.), Institute of Ecology, London.

BALL, I. R., N. GOURBAULT & R. KENK. 1981. The planarians of temporary waters in eastern North America. Contr. Life Sci. R. Ontario Museum 127: 1-27.

BAMBACH, R. K., C. R. SCOTESE & A. M. ZIEGLER. 1980. Before Pangea: the geographies of the Paleozoic World. Amer. Sci. 68: 26-38.

BARBOSA, F. S. & I. BARBOSA. 1958. Dormancy during the larval stages of the trematode *Schistosoma mansoni* in snails aestivating on the soil of dry natural habitats. Ecology 39:763-764.

BARCLAY, M. H. 1966. An ecological study of a temporary pond near Auckland, New Zealand. Aust. J. Mar. Freshwat. Res. 17:239-258.

BARIGOZZI, C. 1980. Genus *Artemia* : problems of systematics. pp. 147-153. *In* : The brine shrimp *Artemia* (G. Persoone, P. Sorgeloos, O. Roels and E. Jaspers, eds.), Vol. 3., Universa Press, Wetteren, Belgium.

BARLOCHER, F., R. J. MACKAY & G. B. WIGGINS. 1978. Detritus processing in a temporary vernal pool in southern Ontario. Arch. Hydrobiol. 81: 269-295.

BARTON, D. R. & S. M. SMITH. 1984. Insects of extremely small and extremely large aquatic habitats. pp. 456-483. *In* : The ecology of aquatic insects (V. H. Resh and D. M. Rosenberg, eds.), Praeger Scientific, N. Y.

BAYLY, I. A. E. 1969. The body fluids of some centropagid copepods: total concentration and amounts of sodium and magnesium. Comp. Biochem. Physiol. 28: 1403-1409.

BAYLY, I. A. E. 1982. Invertebrate fauna and ecology of temporary pools on granite outcrops in Southwestern Australia. Aust. J. Mar. Freshwat. Res. 33: 599-606.

BEADLE, L. C. 1981. The inland waters of tropical Africa. Longman, London. 475pp.

BEAMENT, J. W. L. 1961. The waterproofing mechanism of arthropods II. The permeability of the cuticle of some aquatic insects. J. Exp. Biol. 38: 277-290.

BEAVER, R. A. 1983. The communities living in *Nepenthes* pitcher plants: fauna and food webs. pp. 129-160. *In* : Phytotelmata: terrestrial plants as hosts of aquatic insect communities (J. H. Frank and L. P. Lounibos, eds.), Plexus Publishing, Medford, New Jersey.

BEAVER, R. A. 1985. Geographical variation in food web structure in *Nepenthes* pitcher plants. Ecol. Ent. 10: 241-248.

BEETLE, D. E. 1965. Molluscan fauna of some small ponds in Grand Teton National Park. Nautilus 78: 125-130.

BELCHER, J. H. 1970. The resistance to desiccation and heat of the asexual cysts of some freshwater Prasinophyceae. Br. phycol. 5: 173-177.

BELK, D. 1981. Patterns in anostracan distribution. pp. 168-172. *In* : Vernal pools and intermittent streams (S. Jain and P. Moyle, eds.), Institute of Ecology, Univ. California, Davis, Pub. No. 28.

BENTLEY, P. J. 1966. Adaptations of Amphibia to arid environments. Science 152: 619-623.

BENZING, D. H. 1980. The biology of the bromeliads. Mad River Press, Eureka, California. 305pp.

BERG, K., M. JONASSON &. K.W. OCKELMANN. 1962. The respiration of some animals from the profundal zone of a lake. Hydrobiologia 19: 1-39.

BISHOP, J. A. 1967. Some adaptations of *Limnadia stanleyana* King (Crustacea: Branchiopoda, Conchostraca) to a temporary freshwater environment. J. Anim. Ecol. 36: 599-609.

BISHOP, J. A. 1974. The fauna of temporary rain pools in eastern New South Wales. Hydrobiologia 44: 319-323.

BLANC, M., J. DAGET & F. D'AUBENTON. 1955. Recherches hydrobiologiques dans le bassin du Moyen-Niger. Bull. Inst. Fr. Afr. Noire (A. Sci. Nat.) 17: 619-746.

BONETTO, A. A. 1975. Hydraulic regime of the Parana river and its influence on ecosystems. Ecol. Stud. 10: 175-197.

BONETTO, A. A., C. PIGNELBERI & E. CORDIVIOLA. 1969. Limnological investigations on biotic communities in the Middle Arana River Valley. Verh. int. Verein. theor. angew. Limnol. 17: 1035-1050.

BOROWITZKA, L. J. 1981. The microflora: adaptations to life in extremely saline lakes. pp. 33-46. *In* : Salt lakes (W. D. Williams, ed.), Junk, The Hague.

BOULTON, A. J. & P. J. SUTER. 1986. Ecology of temporary streams - an Australian perspective. pp. 313-327. *In* : Limnology in Australia. (P. De Deckker and W. D. Williams, eds.), CSIRO/Junk Publ., Melbourne and The Netherlands.

BOUVET, Y. 1977. Adaptations physiologiques et compartementales des *Stenophylax* (Limnephilidae) aux eaux temporaires. pp. 117-119. *In* : Proc. Second Int. Symp. on Trichoptera (M. I. Crichton, ed.), Junk, The Hague.

BOWEN, R. 1982. Surface water. Wiley-Interscience, N. Y.

BOWEN, S. T., K. N. HITCHNER & G. L. DANA. 1981. *Artemia* speciation: ecological isolation. pp. 102-114. *In* : Vernal pools and intermittent streams. (S. Jain and P. Moyle, eds.), Institute of Ecology, Univ. California, Davis, Pub. No. 28.

BRAGG, A. N. 1944. Breeding habits, eggs and tadpoles of *Scaphiopus hurterii*. Copeia: 230-241.

BROCH, E. S. 1965. Mechanism of adaptation of the fairy shrimp, *Chirocephalopsis bundyi* Forbes to the temporary pond. Cornell Univ. Agr. Expt. Statn. Mem. 392: 1-48.

BROWN, L. R. & L. H. CARPELAN. 1971. Egg hatching and life history of a fairy shrimp *Branchinecta mackini* Dexter (Crustacea: Anostraca) in a Mohave Desert playa (Rabbit Dry Lake). Ecology 52: 41-54.

BRTEK, J. 1974. Zwei *Streptocephalus* Arten aus Afrika undeinige notizen zur gattung *Streptocephalus*. Annot. Zool. Bot., Bratislava 96: 1-9.

BUCK, J. 1965. Hydration and respiration in chironomid larvae. J. Insect Physiol. 11: 1503-1516.

BURKY, A. J., D. J. HORNBACH & C. M. WAY. 1985. Comparative bioenergetics of permanent and temporary pond populations of the freshwater clam, *Musculium partumeium* (Say). Hydrobiologia 126: 35-48.

BUSHDOSH, M. 1982. Behavioural adaptations to spatially intermittent streams by the longfin dace, *Agosia chrysogaster*, (Cyprinidae). pp. 69-75. *In* : Vernal pools and intermittent streams. (S. Jain and P. Moyle, eds.), Institute of Ecology, Univ. California, Davis, Pub. No. 28.

BUTLER, J. L. 1963. Temperature relations in shallow ponds. Proc. Oklahoma Acad. Sci. 43: 90-95.

BUXTON, P. A. & G. H. E. HOPKINS. 1927. Researches In Polynesia and Melanesia. Mem. London Sch. Trop. Med. Hyg. 1: 1-260.

CALVERT, P. P. 1911. Studies on Costa Rican Odonata. II. The habits of the plant-dwelling larva of *Mecistogaster modestus*.. Ent. News 22: 402-411.

CAMERON, R. E. & G. B. BLANK. 1966. Desert algae: soil crusts and diaphanous substrata as algal habitats. Jet Propulsion Lab., California Inst. Techn., Pasadena, Tech. Rep. 32-971: 1-41.

CAMERON, R. E., F. A. MORELLI & H. P. CONROW. 1970. Survival of microorganisms in desert soil exposed to five years of continuous very high vacuum. Jet Propulsion Lab., California Inst. Techn., Pasadena, Tech. Rep. 32-1454: 1-11.

CAREY, T. G. 1967. Some observations on distribution and abundance of the invertebrate fauna. Fish. Res. Bull., Zambia 3: 22-24.

CARLISLE, D. B. 1968. *Triops* eggs killed only by boiling. Science, N.Y. 161: 279-280.

CARSON, M. A. & E. A. SUTTON. 1971. The hydrologic response of the Eaton River basin, Quebec. Can. J. Earth. Sci. 8: 102-115.

CASPERS, H. & C. W. HECKMAN. 1981. Ecology of orchard drainage ditches along the freshwater section of the Elbe Estuary. Biotic succession and influences of changing agricultural methods. Arch. Hydrobiol. Suppl. 43: 347-486

CASTLE, W. A. 1928. An experimental and histological study of the lifecycle of *Planaria velata*. J. Exp. Biol. 51: 417-476.

CHAMPEAU, A. 1971. Recherches sur l'adaptation de la vie latente des Copepodes Cyclopoides et Harpacticoides des eaux temporaires debasse Provence. Bull. Soc. Ecol. 2: 1-20.

CHENG, L. (ed.). 1976. Marine insects. Elsevier, N.Y. 581pp.

CHIRKOVA, Z. N. 1973. Observations on the survival of cladocerans of the genus *Ilyocryptus* (Macrothricidae) in moist ground. Inf. Byull. Biol. vnutr. Vad. 17: 37-39.

CHODOROWSKI, A. 1969. The desiccation of ephemeral pools and the rate of development of *Aedes communis* larvae. Polskie Archwm Hydrobiol. 16: 79-91.

CLEGG, J. S. 1981. Metabolic consequences of the extent and disposition of the aqueous intracellular environment. J. Exp. Zool. 215: 303-313.

CLIFFORD, H. F. 1966. The ecology of invertebrates in an intermittent stream. Invest. Indiana Lakes Streams 7: 57-98.

COLE, G. A. 1953. Notes on copepod encystment. Ecology 34: 208-211.

COLE, G. A. 1966. Contrasts among calanoid copepods from permanent and temporary ponds in Arizona. Am. Midl. Nat. 76: 351-368.

COLE, G. A. 1968. Desert limnology. pp. 423-486. *In* : Desert Biology (G. W. Brown, ed.), Academic Press, N. Y.

COLE, G. A. & S. G. FISHER. 1979. Nutrient budgets of temporary pond ecosystem. Hydrobiologia 63: 213-222.

COLLINS, N. C. & G. STIRLING. 1980. Relationships among total dissolved solids, conductivity, and osmosity for five *Artemia* habitats (Anostraca: Artemiidae). Grt. Basin Nat. 40: 131-138.

CORBET, P. S. 1963. A biology of dragonflies. Quadrangle Books, Chicago. 247pp.

CORBET, P. S. 1983. Odonata in phytotelmata. pp. 29-54. *In* : Phytotelmata: terrestrial plants as hosts for aquatic insect communities (J. H. Frank and L. P. Lounibos. eds.), Plexus Pub. Inc., Medford, New Jersey.

CROCKER, D. & D. BARR. 1968. Handbook of the crayfish of Ontario. Univ. Toronto Press. 158pp.

CROLL, N. A. 1966. Ecology of parasites. Heinemann, London. 136pp.

CROWE, J. H. 1971. Anhydrobiosis: an unsolved problem. Amer. Naturalist 105: 563-573.

CUMMINS, K. W. 1973. Trophic relations of aquatic insects. Ann. Rev. Ent. 18: 183-206.

DABORN, G. R. 1971. Survival and mortality of coenagrionid nymphs (Odonata: Zygoptera) from the ice of an aestival pond. Can. J. Zool. 49: 69-71.

DABORN, G. R. & H. F. CLIFFORD. 1974. Physical and chemical features of an aestival pond in western Canada. Hydrobiologia 44: 43-59.

DANA, G. L. 1981. *Artemia* in temporary alkaline ponds near Fallon, Nevada with a comparison of its life history strategies in temporary and permanent habitats. pp. 115-125. *In* : Vernal pools and intermittent streams. (S. Jain and P. Moyle, eds.), Institute of Ecology, Univ. California, Davis, Pub. No. 28.

DANCE, K. W. & H. B. N. HYNES. 1979. A continuous study of the drift in adjacent intermittent and permanent streams. Arch. Hydrobiol. 87: 253-261.

DANKS, H. V. 1971. Overwintering of some north temperate and Arctic Chironomidae. II. Chironomid biology. Can. Ent. 103: 1875-1910.

DAVIDSON, J. 1932. Resistance of the eggs of Collembola to drought conditions. Nature, Lond. 29: 867.

DAVIES, R. W., F. J. WRONA & L. LINTON. 1979. A serological study of prey selection by *Helobdella stagnalis* (Hirudinoidea). J. Anim. Ecol. 48: 181-194.

DAVIS, S. N. & R. J. M. DEWIEST. 1966. Hydrogeology. John Wiley and Sons, N.Y. 463 pp.

DECKSBACH, N. K. von. 1929. Zur Klassifikation der Gewasser von astatischen Typus. Arch. Hydrobiol. 20: 349-406.

DONALD, D. B. 1983. Erratic occurrence of anostracans in a temporary pond: colonization and extinction or adaptation to variations in annual weather? Can. J. Zool. 61: 1492-1498.

DRIVER, E. A. 1977. Chironomid communities in small prairie ponds: some characteristics and controls. Freshwat. Biol. 7: 121-133.

DUCHARME, A. 1975. Informe tecnico de biologia pesquera (Limnologia). Publ. Prog. Desarr. Pesca Cont. INDERENA/FAO, Columb., 4: 1-42.

EBERT, T. A. & M. L. BALKO. 1982. Vernal pools as islands in space and time. pp. 90-101. *In* : Vernal pools and intermittent streams. (S. Jain and P. Moyle, eds.), Institute of Ecology, Univ. California, Davis, Pub. No. 28.

ECKBLAD, J. W. 1973. Population studies of three aquatic gastropods in an intermittent backwater. Hydrobiologia 41: 199-219.

EDMONDSON, W. T. (ed.). 1959. Freshwater biology. John Wiley and Sons, N.Y. 1248pp.

EHRENFELD, D. W. 1970. Biological conservation. Holt, Rinehart and Winston, Toronto. 220pp.

ELBORN, C. A. 1966. The life cycle of *Cyclops s. strenuus* Fischer in a small pond. J. Anim. Ecol. 35: 333-347.

ELGMORK, K. 1980. Evolutionary aspects of diapause in freshwater copepods. Spec. Symp. Vol. Am. Soc. Limnol. Oceanogr. 3: 411-417.

ELLIOTT, J. M. 1977. Some methods for the statistical analysis of samples of benthic invertebrates. Freshwat. Biol. Assn. U.K. Sci. Pubn. 25: 1-144.

ERIKSEN, C. H. 1966. Diurnal limnology of two highly turbid puddles. Verh. int. Verein. theor. angew. Limnol 16: 507-514.

EVANS, J. H. 1958. The survival of freshwater algae during dry periods. Part I. Investigation of algae of five small ponds. J. Ecol. 46: 148-167.

EVERITT, D. A. 1981. An ecological study of an Antarctic freshwater pool with particular reference to Tardigrada and Rotifera. Hydrobiologia 83: 225-237.

FALLIS, S. P. & K. R. SNOW. 1983. The hatching stimulus for eggs of *Aedes punctor* (Diptera: Culicidae). Ecol. Ent. 8: 23-28.

FASHING, N. J. 1975. Life history and general biology of *Naiadacarus arboricola* Fashing, a mite inhabiting water-filled treeholes (Acarina: Acaridae). J. nat. Hist. 9: 413-424.

FELSENSTEIN, J. 1976. The theoretical population genetics of variable selection and migration. Ann. Rev. Genet. 10: 253-280.

FELTON, M., J. J. COONEY & W. C. MOORE. 1967. A quantitative study of the bacteria of a temporary pond. J. gen. Microbiol. 47: 25-31.

FENCHEL, T. 1975. The quantitative importance of the benthic microfauna of an Arctic tundra pond. Hydrobiologia 46: 445-464.

FERNANDO, C. H. 1958. The colonization of small freshwater habitats by aquatic insects. I. General discussion, methods and colonization in the aquatic Coleoptera. Ceylon J. Sci.(Bio. Sci.) 1: 117-154.

FERNANDO, C. H. & D. F. GALBRAITH. 1973. Seasonality and dynamics of aquatic insects colonizing small habitats. Verh. int. Verein. theor. angew. Limnol. 18: 1564-1575.

FISCHER, Z. 1961. Some data on the Odonata larvae of small pools. Int. Revue ges. Hydrobiol. 46: 269-275.

FISCHER, Z. 1967. Food composition and food preference in larvae of *Lestes sponsa* (L.) in astatic water environments. Polskie Archwm. Hydrobiol. 14: 59-71.

FISH, D. 1983. Phytotelmata: flora and fauna. pp. 1-28. *In* : Phytotelmata: terrestrial plants as hosts for aquatic insect communities (J. H. Frank and L. P. Lounibos, eds.), Plexus Pub. Inc., Medford, New Jersey.

FISH, D. & D. W. HALL. 1978. Succession and stratification of aquatic insects inhabiting the leaves of the insectivorous pitcher plant, *Sarracenia purpurea*. Am. Midl. Nat. 99: 172-183.

FITTKAU, E. J. *et al.* 1975. Productivity biomass and population dynamics in Amazonian water bodies. Ecol. stud. 11: 289-311.

FOREST, H. S. & C. R. WESTON. 1966. Blue-green algae from the Atacama desert of Northern Chile. J. Phycol. 2: 163-164.

FOX, H. M. 1949. On *Apus* : its rediscovery in Britain, nomenclature and habits. Proc. Zool. Soc. 119: 693-702.

FRANK, J. H. 1983. Bromeliad phytotelmata and their biota, especially mosquitoes. pp. 101-128. *In* : Phytotelmata: terrestrial plants as hosts for aquatic insect communities (J. H. Frank and L. P. Lounibos, eds.), Plexus Pub. Inc., Medford, New Jersey.

FRANK, J. H. & G. A. CURTIS. 1981. Bionomics of the bromeliad-inhabiting mosquito *Wyeomia vanduzeei* and its nursery plant *Tillandsia utriculata.* Florida Ent. 54: 491-506.

FRIEDMANN, E. I. 1964. Xerophytic algae in the Negev Desert. Abstr. 10th Int. Bot. Congr. pp. 290-291.

FRIEDMANN, E. I., Y. LIPKIN & R. OCAMPO-PAUS. 1967. Desert algae of the Negev (Israel). Phycologia 6: 185-196.

FRITH, H. J. 1959. Ecology of wild ducks in inland Australia. pp. 383-395. *In* : Biogeography and ecology in Australia (A. Keast, R. L. Crocker and C. S. Christian, eds.), Monographiae Biologicae 3, Junk, The Hague.

FRYER, G. 1974. Attachment of bivalve molluscs to corixid bugs. Naturalist, Hull. 28: 18.

GANNING, B. 1971. Studies on chemical, physical and biological conditions in Swedish rockpool ecosystems. Ophelia 9: 51-105.

GAUSE, G. F. 1934. The struggle for existence. Williams Wilkins, Baltimore, Hafner, N.Y. 163pp. (reprinted 1964).

GEDDES, M. C. 1976. Seasonal fauna of some ephemeral saline waters in western Victoria with particular reference to *Parartemia zietziana* Sayce (Crustacea: Anostraca). Aust. J. Mar. Freshwat. Res. 27: 1-22.
GEDDES, M. C. 1980. The brine shrimps *Artemia* and *Parartemia* in Australia. pp. 57-65. *In* : The brine shrimp *Artemia* (G. Persoone, P. Sargeloos, D. Roels and E. Jaspers, eds.), Vol. 3. Universa Press, Wetteren, Belgium.
GEKKO, K. & S. N. TIMASHEFF. 1981. Thermodynamic and kinetic examination of protein stabilization by glycerol. Biochemistry 20: 4677-4686.
GEORGE, M. G. 1961. Diurnal variations in two shallow ponds in Delhi, India. Hydrobiologia 18: 265-273.
GIESEL, J. T. 1976. Reproductive strategies as adaptations to life in temporally heterogeneous environments. Ann. Rev. Ecol. Syst. 7: 57-79.
GONZALEZ-JIMEREZ, E. 1977. The capybara. World Anim. Rev. 24-30.
GRANT, I. F. & R. SEEGERS. 1985. Tubificid role in soil mineralization and recovery of algal nitrogen by lowland rice. Soil Biol. Biochem. 17: 559-563.
GRENSTED, L. W. 1939. Colonization of new areas by water beetles. Entomologist's mon. Mag. 75: 174-175.
GRIMM, N. B., S. G. FISHER & W. L. MINCKLEY. 1981. Nitrogen and phosphorus dynamics in hot desert streams of southwestern U.S.A. Hydrobiologia 83: 303-312.
GROW, L. & H. MERCHANT. 1980. The burrow habitat of the crayfish *Cambarus diogenes diogenes* (Girard). Am. Midl. Nat. 103: 231-237.
GUBLER, D. J. 1971. Studies on the comparative oviposition behaviour of *Aedes (Stegomyia) albopictus* and *Aedes (S.) polynesiensis* Marks. J. Med. Ent. 8: 675-682.
HALL, F. G. 1922. The vital limit of desiccation of certain animals. Biol. Bull. Woods Hole 42: 31-51.
HAMMER, O. 1941. Biological and ecological investigations on flies associated with pasturing cattle and their excrement. Vidensk. Medd. dansk. naturh. Foren. Kbh. 105: 1-257.
HARPER, P. P. & H. B. N. HYNES. 1970. Diapause in the nymphs of Canadian winter stoneflies. Ecology 51: 425-427.
HARRIS, R. E. & W. A. G. CHARLESTON. 1977. The response of the freshwater gastropods *Lymnaea tomentosa* and *L. columella* to desiccation. J. Zool. 183: 41-46.
HARRISON, L. 1922. On the breeding habits of some Australian frogs. Aust. Zool. 3: 17-34.
HARTLAND-ROWE, R. 1972. The limnology of temporary waters and the ecology of the Euphyllopoda. pp. 15-31. *In* : Essays in hydrobiology (R. B. Clark and R. J. Wootton, eds.), Exeter Univ. Press, Exeter. 316pp.
HENRY, W. A. & F. E. MORRISON. 1923. Foods and feeding. Madison.
HILDREW, A. G. 1985. A quantitative study of the life history of fairy shrimp (Branchiopoda: Anostraca) in relation to the temporary nature of its habitat, a Kenyan rainpool. J. Anim. Ecol. 54: 99-110.
HINTON, H. E. 1951. A new chironomid from Africa, the larva of which can be dehydrated without injury. Proc. zool. Soc.Lond. 121: 371-380.
HINTON, H. E. 1952. Survival of chironomid larva after twenty months dehydration. Trans. 9th Int. Congr. Ent. 1: 478-482.
HINTON, H. E. 1953. Some adaptations of insects to environments that are alternately dry and flooded, with some notes on the habits of the Stratiomyidae. Trans. Soc. Br. Ent. 11: 209-227.
HINTON, H. E. 1954. Resistance of the dry eggs of *Artemia salina* L. to high temperature. Ann. Mag. nat. Hist. 7: 158-160.
HOCHACHKA, P. W. & G. N. SOMERO. 1984. Biochemical adaptation. Princeton Univ. Press, New Jersey. 537pp.

HOHAM, R. W., J. E. ULLET & S. C. ROEMER. 1983. The life history and ecology of the snow alga *Chloromonas polyptera,* new combination Chlorophyta: Volvocales. Can. J. Bot. 61: 2416-2429.

HOLLAND, R. F. & S. K. JAIN. 1981. Spatial and temporal variation in plant species diversity of vernal pools. pp. 198-209. *In* : Vernal pools and intermittent streams (S. Jain and P. Moyle, eds.), Institute of Ecology, Univ. California, Davis, Pub. No. 28.

HORSFALL, W. R. 1955. Mosquitoes, their bionomics and relation to disease. Ronald Press, N.Y. 723pp.

HORSFALL, W. R. 1956. Eggs of floodwater mosquitoes. III. Conditioning and hatching of *Aedes vexans* (Diptera: Culicidae). Ann. ent. Soc. Am. 49: 66-71.

HORTON, R. E. 1933. The role of infiltration in the hydrogeologic cycle. Am. Geophys. Union Trans. 14: 446-460.

HOWARD-WILLIAMS, C. & G. M. LENTON. 1975. The role of the littoral zone in the functioning of a shallow tropical lake ecosystem. Freshwat. Biol. 5: 445-459.

HUNT, P. C. & J. W. JONES. 1972. The effect of water level fluctuations on a littoral fauna. J. Fish. Biol. 4: 385-394.

HUSBY, J. A. & K. E. ZACHARIASSEN. 1980. Antifreeze agents in the body fluid of winter active insects and spiders. Experientia 36: 963-964.

HYNES, H. B. N. 1961. The effect of water-level fluctuation on littoral fauna. Verh. int. Verein. theor. angew. Limnol. 14: 652-655.

IMHOFF, J. G. A. & A. D. HARRISON. 1981. Survival of *Diplectrona modesta* Banks (Trichoptera: Hydropsychidae) during short periods of desiccation. Hydrobiologia 77: 61-63.

ISTOCK, C. A., K. J. VAVRA & H. ZIMMER. 1976. Ecology and evolution of the pitcher-plant mosquito. 3. Resource tracking by a natural population. Evolution 30: 548-557.

ITAMIES, J. & E. LINDGREN. 1985. The ecology of *Chionea* species (Diptera, Tipulidae). Notulae Ent. 65: 29-31.

JACKSON, D. J. 1956. Observations on water beetles during drought. Entomologist's mon. Mag. 92: 154-155.

JAMES, H. G. 1969. Immature stages of five diving beetles (Coleoptera: Dytiscidae); notes on their habits and life histories, and a key to aquatic beetles of vernal woodland pools in southern Ontario. Proc. ent. Soc. Ont. 100: 52-97.

JONES, A. W. 1967. Introduction to parasitology. Addison-Wesley, London. 458pp.

JONES, J. R. E. 1949. A further ecological study of calcareous streams in the Black Mountains district of South Wales. J. Anim. Ecol. 18: 142-159.

JONES, R. E. 1975. Dehydration in an Australian rockpool chironomid larva *(Paraborniella tonnoiri).* J. Ent. (A) 49: 111-119.

JORDAN, P. & G. WEBBE. 1969. Human schistosomiasis. Heinemann, London. 212pp.

KAISILA, J. 1952. Insects from arctic mountain snows. Ann. Ent. Fenn. 18: 8-25.

KAPLAN, R. H. 1981. Temporal heterogeneity of habitats in relation to amphibian ecology. pp. 143-154. *In* : Vernal pools and intermittent streams (S. Jain and P. Moyle, eds.), Institute of Ecology, Univ. California, Davis, Pub. No. 28.

KASTER, J. L. & J. H. BUSHNELL. 1981. Cyst formation by *Tubifex tubifex* (Tubificidae). Trans. Am. microscop. Soc. 100: 34-41.

KEILIN, D. 1944. Respiratory systems and respiratory adaptations in larvae and pupae of Diptera. Parasitology 36: 1-36.

KENK, R. 1944. The freshwater triclads of Michigan. Misc. Publs. Mus. Zool. Univ. Mich. 60: 1-44.

KENK, R. 1949. The animal life of temporary and permanent ponds in southern Michigan. Misc. Publs. Mus. Zool. Univ. Mich. 71: 1-66.

KERFOOT, W. C. 1978. Combat between predatory copepods and their prey: *Cyclops, Epischura* and *Bosmina*. Limnol. Oceanogr. 23: 1089-1103.

KINGSOLVER, J. G. 1979. Thermal and hydric aspects of environmental heterogeneity in the pitcher plant mosquito. Ecol. Monogr. 49: 357-376.

KITCHING, R. L. 1971. An ecological study of water-filled treeholes and their position in the woodland ecosystem. J. Anim. Ecol. 40: 281-302.

KITCHING, R. L. 1983. Community structure in water-filled treeholes in Europe and Australia - comparisons and speculations. pp. 205-222. *In* : Phytotelmata: terrestrial plants as hosts for aquatic insect communities (J. H. Frank and L. P. Lounibos, eds.), Plexus Pub. Inc., Medford, New Jersey.

KLIMOWICZ, H. 1959. Tentative classification of small water bodies on the basis of the differentiation of the molluscan fauna. Polskie Archwm. Hydrobiol. 6: 85-104.

KNOWLTON, G. F. 1951. A flight of water boatmen. Bull. Brooklyn ent. Soc. 46: 22-23.

KUSHLAN, J. A. 1973. Differential responses to drought in two species of Fundulus. Copeia: 808-809.

KUSHNER, D. J. 1978. Life in high salt and solute concentrations: halophilic bacteria. pp. 317-368. *In* : Microbial life in extreme environments (E. J. Kushner, ed.), Academic Press, London.

LAINSON, R. & J. J. SHAW. 1971. Epidemiological considerations of the leishmanias with particular reference to the New World. pp. 21-57. *In* : Ecology and physiology of parasites (A.M. Fallis, ed.), Univ. Toronto Press, Toronto.

LAKE, P. S. 1977. Pholeteros, the faunal assemblage found in crayfish burrows. Newsl. Aust. Soc. Limnol. 15: 57-60.

LAKE, P. S. 1982. Ecology of the macroinvertebrates of Australian upland streams- a review of current knowledge. Bull. Aust. Soc. Limnol. 8: 1-15.

LANDIN, J. 1968. Weather and diurnal periodicity of flight by *Helophorus brevipalpis* Bedel (Coleoptera: Hydrophilidae). Opusc. ent. 33: 28-36.

LANDIN, J. 1980. Habitats, life histories, migration and dispersal by flight of two water beetles *Helophorus brevipalpis* and *H. strigifrons* (Hydroptilidae). Holarct. Ecol. 3: 190-201.

LANDIN, J. & E. STARK. 1973. On flight thresholds for temperature and wind velocity, 24-hour flight periodicity and migration of the water beetle *Helophorus brevipalpis* Bedel. Zoon (suppl.)1: 105-114.

LARSON, D. J. 1985. Structure in temperate predaceous diving beetle communities (Coleoptera: Dytiscidae). Holarct. Ecol. 8: 18-32.

LAURENCE, B. R. 1954. The larval inhabitants of cow pats. J. Anim. Ecol. 23: 234-260.

LEADER, J. P. 1962. Tolerance to freezing of hydrated and partially hydrated larvae of *Polypedilum* (Chironomidae). J. Insect Physiol. 8: 155-163.

LEES, A. D. 1953. The physiology of diapause in arthropods. Cambridge Univ. Press. 151pp.

LEEWENHOEK, A. 1701. *cited in* Hall, F. G. 1922. The vital limit of exsiccation of certain animals. Biol. Bull. Woods Hole 42: 31-51.

LEHMKUHL, D. M. 1973. A new species of *Baetis* from ponds in the Canadian Arctic, with biological notes. Can. Ent. 10: 343-346.

LEINAAS, H. P. 1981a. Activity of Arthropoda in snow within coniferous forest, with special reference to Collembola. Holarct. Ecol. 4: 127-138.

LEINAAS, H. P. 1981b. Cyclomorphosis in the furca of the winter active Collembola *Hypogastrura socialis* (Uzel). Ent. scand. 12: 35-38.

LEOPOLD, L. B., M. B. WOLMAN & J. P. MILLER. 1964. Fluvial processes in geomorphology. W. H. Freeman, San Francisco. 522pp.

LEWIS, M. A. & S. D. GERKING. 1979. Primary productivity in a polluted intermittent desert stream. Am. Midl. Nat. 102: 172-174.
LEWONTIN, R. C. 1964. Selection for colonizing ability. *In* : The genetics of colonizing species (H. G. Baker, ed.), Academic Press, N.Y. 588pp.
LICHTI-FEDEROVICH, S. 1980. Diatom flora of red snow from Isbjorneo, Carey Oer, Greenland. Nova Hedw. 33: 395-420.
LINDER, F. 1959. Notostraca. pp. 572-576. *In* : Freshwater biology (2nd ed.) (W. T. Edmondson, ed.), John Wiley and Sons, N.Y.
LIPPERT, B. E. & D. L. JAMESON. 1964. Plant succession in temporary ponds at Willamette Valley, Oregon. Am. Midl. Nat. 71: 181-197.
LOUNIBOS, L. P. 1980. The bionomics of three sympatric *Eretmapodites* (Diptera: Culicidae) at the Kenya coast. Bull. Ent. Res. 70: 309-320.
LOUNIBOS, L. P., C. V. DOVER & G. F. O'MEARA. 1982. Fecundity, autogeny, and the larval environment of the pitcher-plant mosquito, *Wyeomyia smithii* . Oecologia 55: 160-164.
LOWE-McCONNELL, R. H. 1977. Ecology of fishes in tropical waters. Edward Arnold, London. 64pp.
MACAN, T. T. 1939. Notes on the migration of some aquatic insects. J. Soc. Br. Ent. 2: 1-6.
MACARTHUR, R. H. & E. O. WILSON. 1963. An equilibrium theory of insular zoogeography. Evolution. 17: 373-387.
MACARTHUR, R. H. & E. O. WILSON. 1967. The theory of island biogeography. Princeton, University Press, New Jersey. 203pp.
McLACHLAN, A. J. 1981. Food resources and foraging tactics in tropical rain pools. Zool. J. Linn. Soc. 71: 265-277.
McLACHLAN, A. J. 1983. Life-history tactics of rain-pool dwellers. J. Anim. Ecol. 52: 545-561.
McLACHLAN, A. J. 1985. What determines the species present in a rain-pool? Oikos 45: 1-7.
McLACHLAN, A. J. 1986. Sexual dimorphism in midges: strategies for flight in the rain-pool dweller *Chironomus imicola* (Diptera: Chironomidae). J. Anim. Ecol. 55: 261-267.
McLACHLAN, A. J. & M. A. CANTRELL. 1980. Survival strategies in tropical rain pools. Oecologia 47: 344-351.
McLACHLAN, A. J., P. R. MORGAN, C. HOWARD-WILLIAMS, S. M. McLACHLAN, & D. BOURN. 1972. Aspects of the recovery of a saline African lake following a dry period. Arch. Hydrobiol. 70: 325-340.
MACHADO-ALLISON, D. J. RODRIGUEZ, R. BERRERA & C. G. COVA. 1983. The insect community associated with inflorescences of *Heliconia caribaea* Lamark, in Venezuela. pp. 247-270. *In* : Phytotelmata: terrestrial plants as hosts for aquatic insect communities (J. H. Frank and L. P. Lounibos, eds.), Plexus Pub. Inc., Medford, New Jersey.
MAGUIRE, B. 1963. The passive dispersal of small aquatic organisms and their colonization of isolated bodies of water. Ecol. Monogr. 33: 161-185.
MAIN, A. R., M. J. LITTLEJOHN & A. K. LEE. 1959. Ecology of Australian frogs. pp. 398-411. *In* : Biogeography and ecology in Australia (A. Keast, R. L. Crocker and C. S. Christian, eds.), Monographiae Biologicae 3, Junk, The Hague.
MALY, E. J., S. SCHAENHOLTZ & M. T. ARTS. 1980. The influence of flatworm predation on zooplankton inhabiting small ponds. Hydrobiologia 76: 233-240.
MARCHANT, H. J. 1982. Snow algae from the Australian snowy mountains. Phycologia 21: 178-184.
MARCHANT, R. 1982a. Seasonal variation in the macroinvertebrate fauna of Billabongs along Magela Creek, Northern Territory. Aust. J. Mar. Freshwat. Res. 33: 329-342.

MARCHANT, R. 1982b. Life spans of two species of tropical mayfly nymph (Ephemeroptera) from Magela Creek, Northern Territory. Aust. J. Mar. Freshwat. Res. 33: 173-179.

MATTINGLY, P. F. 1969. The biology of mosquito-borne diseases. George Allen and Unwin, London. 184pp.

MATTOX, N. T. 1959. Conchostraca. pp. 577-586. *In* : Freshwater biology (2nd ed.) (W. T. Edmondson, ed.), John Wiley and Sons, N.Y.

MAWSON, D. 1950. Occurrence of water in Lake Eyre, South Australia. Nature, Lond. 166: 667-668.

MAYHEW, W. H. 1965. Adaptations of the amphibian *Scaphiopus couchii* to desert conditions. Amer. Midl. Nat. 74: 95-109.

MAYHEW, W. H. 1968. Biology of desert amphibians and reptiles. pp. 195-356. *In*: Desert biology, Vol. 1 (G. W. Brown, ed.), Academic Press, N.Y.

MELLOR, M. W. 1979. A study of the salt lake snail *Coxiella* (Smith) 1844, *sensu lato*. B.Sc. (Hons) thesis, Univ. Adelaide. 67pp.

MENGE, B. A. & J. P. SUTHERLAND. 1976. Species diversity gradients: synthesis of the roles of predation, competition, and temporal heterogeneity. Amer. Nat. 110: 350-369.

MINCKLEY, W. L. & W. E. BARBER. 1979. Some aspects of the biology of the longfin dace, a cyprinid fish characteristic of streams in the Sonoran desert. Southwest. Nat. 15: 459-464.

MOGHRABY, A. I. 1977. A study on diapause of zooplankton in a tropical river - The Blue Nile. Freshwat. Biol. 7: 77-117.

MONAKOV, A. V. 1969. The zooplankton and the zoobenthos of the White Nile and adjoining waters in the Republic of Sudan. Hydrobiologia 33: 161-185.

MOORE, W. G. & A. BURN. 1968. Lethal oxygen thresholds for certain temporary pond invertebrates and their applicability to field situations. Ecology 49: 349-351.

MORLEY, A. W., T. E. BROWN & D. V. KOONTZ. 1985. The limnology of a naturally acidic tropical water system in Australia I. General description and wet season characteristics. Verh. int. Verein. theor. angew. Limnol. 22: 2125-2130.

MORTON, D. W. & I. A. E. BAYLY. 1977. Studies on the ecology of some temporary freshwater pools in Victoria with special reference to microcrustaceans. Aust. J. Mar. Freshwat. Res. 28: 434-454.

MOZLEY, A. 1944. Temporary ponds, neglected natural resource. Nature, Lond. 154: 490.

MULLER, F. 1980. Wasserthiere in Baumwipfeln. Kosmos (Stockholm). 6: 386-388.

NILSSON, A. N. 1986. Community structure in the Dytiscidae (Coleoptera) of a northern Swedish seasonal pond. Ann. Zool. Fennici 23: 39-47.

ODUM, E. P. 1971. Fundamentals of ecology. W. B. Saunders Co., London. 574pp.

OLSEN, O. W. 1974. Animal parasites: their life cycles and ecology. University Park Press, Baltimore. 562pp.

OLSSON, T. I. 1981. Overwintering of benthic macroinvertebrates in ice and frozen sediment in a North Swedish River. Holarct. Ecol. 4: 161-166.

OMER, S. M. & J. L. CLOUDSLEY-THOMPSON. 1968. Dry season biology of *Anopheles gambiae* Giles in the Sudan. Nature, Lond. 217: 879-880.

OTTO, C. 1976. Habitat relationships in the larvae of three Trichoptera species. Arch. Hydrobiol. 77: 505-517.

OUTRIDGE, P. 1986. High species diversity in tropical Australian freshwater macrobenthic community. Hydrobiologia (In press).

PAJOT, F. 1983. Phytotelmata and mosquito vectors of sylvatic yellow fever in Africa. pp. 79-100. *In* : Phytotelmata: terrestrial plants as hosts for aquatic insect communities (J. H. Frank and L. P. Lounibos, eds.), Plexus Pub. Inc., Medford, New Jersey.

PAJUNEN, V. & A. JANSSON. 1969. Dispersal of rock-pool corixids *Arctocorixa carinata* (Sahler) and *Callicorixa producta* (Reut) (Heteroptera: Corixidae). Ann. Zool. Fenn. 6: 391-427.

PARKER, B. C., N. SCHANEN & R. RENNER. 1969. Viable soil algae from the herbarium of the Missouri Botanical Gardens. Ann. Mo. Bot. Gard. 56: 113-119.

PATERSON, C. G. & C. J. CAMERON. 1982. Seasonal dynamics and ecological strategies of the pitcher plant chironomid *Metriocnemus knabi* Coq. (Diptera: Chironomidae) in southeastern New Brunswick. Can. J. Zool. 60: 3075-3083.

PEARCE, E. J. 1939. Colonization by water beetles. Entomologist's mon. Mag. 75: 208.

PENNAK, R. W. 1953. Freshwater invertebrates of the United States. Ronald Press, N.Y. 769pp.

PIANKA, E. R. 1970. On *r*- and K-selection. Am. Nat. 592-597.

PICHLER, W. 1939. Unsere derzeitige Kenntnis van der Thermik Kleiner Gewasser Thermische Kleingewassertypen. Int. Revue ges. Hydrobiol. 38: 231-242.

POPHAM, E. J. 1953. Observations on the migration of corixids (Hemiptera) into a new aquatic habitat. Entomologist's mon. Mag. 89: 124-125.

PRICE, P. W. 1984. Alternative paradigms in community ecology. pp. 353-383. *In* : A new ecology. Novel approaches to interactive systems. (P. W. Price, C. N. Slobodchikoff and S. Gaud, eds.), John Wiley and Sons, N. Y.

PROCTOR, V. W. 1964. Viability of crustacean eggs recovered from ducks. Ecology 45: 656-658.

RACKHAM, O. 1986. The history of the countryside. J. M. Dent and Sons, London. 445pp.

REID, G. K. 1961. Ecology of inland waters and estuaries. Van Nostrand Reinholt, N.Y. 375pp.

REISS, F. 1977. Qualitative and quantitative investigations on the macrobenthic fauna of Central Amazon lakes. I. Lago Tupe, a black water lake on the lower Rio Negro. Amazoniana 6: 203-235.

REX, R. W. 1961. Hydrodynamical analysis of circulation and orientation of lakes in northern Alaska. pp. 1021-1043. *In* : G. O. Raasch (ed.), Geology of the Arctic. Univ. Toronto Press.

RICHARDS, C. S. 1967. Aestivation of *Biomphalaria glabrata* (Basommatophora: Planorbidae). Amer. J. Trop. Med. Hygiene 16: 797-802.

ROHNERT, U. 1950. Wasserfullte Baumhohlen und ihre Besiedlung. Ein Beitrag zur Fauna Dendrolimnetica. Arch. Hydrobiol. 44: 472-516.

RZOSKA, J. 1974. The Upper Nile swamps, tropical wetland study. Freshwat. Biol. 4: 1-30.

RZOSKA, J. 1984. Temporary and other waters. *In* : Sahara Desert (J. L. Cloudsley-Thompson, ed.), Pergamon Press, Oxford. 348pp.

SANDS, A. 1981. Algae of vernal pools and intermittent streams. pp. 66-68. *In:* Vernal pools and intermittent streams. (S. Jain and P. Moyle, eds.), Institute of Ecology, Univ. California, Davis, Pub. No. 28.

SANDERS, H. L. 1968. Marine benthic diversity: comparative study. Amer. Nat. 102: 243-282.

SCHMITT, W. L. 1971. Crustaceans. Univ. Michigan Press, Ann Arbor. 204pp.

SCHNELLER, M. V. 1955. Oxygen depletion in Salt Creek, Indiana. Invest. Indiana Lakes and Streams 4: 163-175.

SCHOOF, H. F., S. C. SCHELL & D. F. ASHTON. 1945. Survival of anopheline larvae and pupae in muck. J. econ. Ent. 38: 113-114.

SCHWABE, G. H. 1963. Blaualgen der phototropen Grengschicht. Blaualgen und Lebensraum. VII. Pedobiologia 2: 132-152.

SCOURFIELD, D. J. & J. P. HARDING. 1966. A key to the British species of freshwater Cladocera. Freshwater Biological Assn. U.K., Sci. Pubn. 5: 1-55.

SERVICE, M. W. 1977. Ecological and biological studies on *Aedes cantans* (Meig.) (Diptera: Culicidae) in southern England. J. Appl. Ecol. 14: 159-196.

SHEATH, R. G. & J. A. HELLEBUST. 1978. Comparison of algae in the euplankton, tychoplankton and periphyton of a tundra pond. Can. J. Bot. 56: 1472-1483.

SIOLI, H. 1975. Amazon tributaries and drainage basins. Ecol. stud. 10: 199-213.

SMITH, L. B. 1953. Bromeliad malaria. Ann. Rep. Smithsonian Instit. (1952) 385-398.

SMITH, R. E. W. 1983. Community dynamics of the pool fauna in an intermittent stream. Austr. soc. Limnol. Newsletter 21: 18.

SOMME, L. 1964. Effects of glycerol on cold-hardiness in insects. Can. J. Zool. 42: 87-101.

STEARNS, S. C. 1976. Life-history tactics: review of the ideas. Quart. Rev. Biol. 51: 3-47.

STOCKER, Z. S. J. & H. B. N. HYNES. 1976. Studies on the tributaries of Char Lake, Cornwallis Island, Canada. Hydrobiologia 49: 97-102.

STOUT, V. M. 1964. Studies in temporary ponds in Canterbury, New Zealand. Verh. int. Verein. theor. angew. Limnol. 15: 209-214.

STRANDINE, E. J. 1941. Effect of soil moisture and algae on the survival of a pond snail during periods of relative dryness. Nautilus 54: 128-130.

STRONG, D. R., D. SIMBERLOFF, L. G. ABELE & A. B. THISTLE. 1984. Ecological communities, conceptual issues and the evidence. Princeton University Press, Princeton, New Jersey. 613pp.

STURROCK, R. F. 1974. Ecological notes on habitats of the freshwater snail *Biomphalaria glabrata* , intermediate host of *Schistosoma mansoni* on St. Lucia, West Indies. Carib. J. Sci. 14: 149-162.

STYCZYNSKA-JUREWICZ, E. 1966. Astatic-water bodies as characteristic habitat of some parasites of man and animals. Verh. int. Verein. theor. angew. Limnol. 16: 604-611.

SUGDEN, A. M. & R. J. ROBINS. 1979. Aspects of the ecology of vascular epiphytes in Colombian closed forests I. The distribution of the epiphytic flora. Biotropica 11: 173-188.

SUTCLIFFE, D. W. 1960. Osmotic regulation in the larvae of some euryhaline Diptera. Nature, Lond. 187: 331-332.

SUTCLIFFE, D. W. 1961. Salinity fluctuations and the fauna in a salt marsh with special reference to aquatic insects. Trans. nat. Hist. Soc. Northumb. Durh. 14: 37-56.

TALLING, J. F. 1951. The element of chance in pond population. Naturalist 4: 157.

TASCH, P. 1969. Branchiopoda. pp. 128-191. *In* : Treatise on invertebrate paleontology (R. C. Moore, ed.), Part R (4), Vol.1. Geological Society of America, Boulder, Colorado.

TAYLOR, C. C. 1972. Medieval moats in Cambridgeshire pp. 237-249. *In* : P. J. Fowler (ed.), Archeology and the landscape. J. Barker. London.

TEMPLETON, A. R. 1980. Modes of speciation and inferences based on genetic distances. Evolution 34: 719-729.

THOMAS, G. I. 1963. Study of population of sphaeriid clams in a temporary pond. Nautilus 77: 37-43.

TOWNS, D. R. 1985. Limnological characteristics of South Australian intermittent stream, Brownhill Creek. Aust. J. Mar. Freshw. Res. 36: 821-837.

TRAINOR, F. R. 1962. Temperature tolerance of algae in dry soil. Phycol. Soc. Amer., News Bull. 15: 3-4.

TRUCHELUT, A. 1982. Coutumes et usages des etangs de la Dombes et de la Bresse (par C. Rivoire). Editions de Trevoux. 170pp.

VASS, K. F. & M. SACHLAN. 1955. Limnological studies on diurnal fluctuations in shallow ponds in Indonesia. Verh. int. Verein. theor. angew. Limnol. 12: 309-319.

VICTOR, R. & C. H. FERNANDO. 1978. Systematics and ecological notes on Ostracoda from container habitats of some South Pacific Islands. Can. J. Zool. 56: 414-422.

VINCENT, W. F. & C. HOWARD-WILLIAMS. 1986. Antarctic stream ecosystems: physiological ecology of a blue-green algal epilithon. Freshwat. Biol. 16: 209-233.

VOGEL, S. 1955. Niedere Fensterpflanzen in der sudafrikanischen Wuste Bietr. Biol. Pflanz. 31: 45-135.

WALKER, T. D. & P. A. TYLER. 1979. A limnological survey of the Magela Creek system, Alligator Rivers region. Northern Territory. First and Second Interim Rept. Sup. Scientist, Alligator Rivers region, Sydney.

WARD, J. V. & J. A. STANFORD (eds.) 1979. The ecology of regulated streams. Plenum Publ. N.Y., 398pp.

WATTS, R. B. & S. M. SMITH. 1978. Oogenesis in *Toxorhynchites rutulis* (Diptera: Culicidae). Can. J. Zool. 56: 136-139.

WAY, C. M., D. J. HORNBACH & A. J. BURKY. 1980. Comparative life history tactics of the sphaeriid clam, *Musculium partumeium* (Say), from a permanent and temporary pond. Am. Midl. Nat. 104: 319-327.

WEAVER, C. R. 1943. Observations on the life history of the fairy shrimp *Eubranchipus vernalis* . Ecology 24: 500-502.

WEISS, R. L. 1983. Fine structure of the snow alga *Chlamydomonas nivalis* and associated bacteria. J. Phycol. 19: 200-204.

WELCOMME, R. L. 1979. Fisheries ecology of floodplain rivers. Longman, London. 317pp.

WHITNEY, R. J. 1942. Diurnal fluctuations of oxygen and pH in two small ponds and a stream. J. Exp. Biol. 19: 92-99.

WIGGINS, G. B. 1973. A contribution to the biology of caddisflies (Trichoptera) in temporary pools. Life Sci. Contr. R. Ontario Museum 88: 1-28.

WIGGINS, G. B. & R. J. MACKAY. 1978. Some relationships between systematics and trophic ecology in nearctic aquatic insects, with special reference to Trichoptera. Ecology 59: 1211-1220.

WIGGINS, G. B., R. J. MACKAY & I. M. SMITH. 1980. Evolutionary and ecological strategies of animals in annual temporary pools. Arch. Hydrobiol. (suppl.) 58: 97-206.

WILBUR, H. M. & J. P. COLLINS. 1973. Ecological aspects of amphibian metamorphosis. Science 182: 1305-1314.

WILLIAMS, D. D. 1980. Temporal patterns in recolonization of stream benthos. Arch. Hydrobiol. 90: 56-74.

WILLIAMS, D. D. 1983. The natural history of nearctic temporary pond in Ontario with remarks on continental variation in such habitats. Int. Revue ges. Hydrobiol. 68: 239-253.

WILLIAMS, D. D. & B. W. COAD. 1979. The ecology of temporary streams III. Temporary stream fishes in southern Ontario, Canada. Int. Revue ges. Hydrobiol. 64: 501-515.

WILLIAMS, D. D. & H. B. N. HYNES. 1976. The ecology of temporary streams I. The faunas of two Canadian streams. Int. Revue ges. Hydrobiol. 61: 761-787.

WILLIAMS, D. D. & H. B. N. HYNES. 1977a. Benthic community development in a new stream. Can. J. Zool. 55: 1071-1076.

WILLIAMS, D. D. & H. B. N. HYNES. 1977b. The ecology of temporary streams II. General remarks on temporary streams. Int. Revue ges. Hydrobiol. 62: 53-61.

WILLIAMS, D. D. & N. E. WILLIAMS. 1975. A contribution to the biology of *Ironoquia punctatissima* (Trichoptera: Limnephilidae). Can. Ent. 107: 829-832.

WILLIAMS, D. D. & N. E. WILLIAMS. 1976. Aspects of the ecology of the faunas of some brackishwater pools on the St. Lawrence North Shore. Can. Field-Nat. 90: 410-415.

WILLIAMS, D. D., N. E. WILLIAMS & H. B. N. HYNES. 1974. Observations on the life history and burrow construction of the crayfish *Cambarus fodiens* (Cottle) in a temporary stream in southern Ontario. Can. J. Zool. 52: 365-370.

WILLIAMS, W. D. 1968. The distribution of *Triops* and *Lepidurus* (Branchiopoda) in Australia. Crustaceana 14:119-126.

WILLIAMS, W. D. 1975. A note on the macrofauna of a temporary rain pool in semi-arid Western Australia. Aust. J. Mar. Freshwat. Res. 26: 425-429

WILLIAMS, W. D. 1980. Australian freshwater life. Macmillan, London. 321pp.

WILLIAMS, W. D. 1981a. Running water ecology in Australia. pp.367-392. *In*: Perspectives in running water ecology (M. A. Lock and D. D. Williams, eds.), Plenum, N.Y.

WILLIAMS, W. D. (ed.) 1981b. Salt lakes. Junk, The Hague.

WILLIAMS, W. D. 1983. Life in inland waters. Blackwells Sci. Pub., Oxford. 252pp.

WILLIAMS, W. D. 1984. Chemical and biological features of salt lakes on the Eyre Peninsula, South Australia, and an explanation of regional differences in the fauna of Australian salt lakes. Verh. int. Verein. theor. angew. Limnol. 22: 1208-1215.

WILLIAMS, W. D. 1985. Biotic adaptations in temporary lentic waters with special reference to those in semi-arid regions. pp.85-110. *In*: Perspectives in Southern Hemisphere Limnology (B. R. Davies and R. D. Walmsley, eds.), Junk, The Hague.

WILLIAMS, W. D. & R. T. BUCKNEY. 1976. Stability of ionic proportions in five salt lakes in Victoria, Australia. Aust. J. Mar. Freshwat. Res. 27: 367-377.

WINFIELD, T. P., T. CASS & K. B. MACDONALD. 1981. Small mammal utilization of vernal pools, San Diego County, California. pp. 161-167. *In*: Vernal pools and intermittent streams (S. Jain and P. Moyle, eds.), Institute of Ecology, Univ. California, Davis, Pub. No. 28.

WINTERBOURN, M. J. & N. H. ANDERSON. 1980. The life history of *Philanisus plebeius* Walker (Trichoptera: Chathamiidae), a caddisfly whose eggs were found in a starfish. Ecol. Ent. 5: 293-303.

WOLVERTON, B. & R. C. McDONALD. 1976. Don't waste waterweeds. New Sci. 71: 318-320.

WOODWARD, D. B., H. C. CHAPMAN & J. J. PETERSEN. 1968. Laboratory studies on the seasonal hatchability of egg batches of *Aedes sollicitans*, *A. taeniorhynchus* and *Psorophora confinnis*. Mosquito News. 28: 143-146.

WRATTEN, S. D. & G. L. A. FRY. 1980. Field and laboratory exercises in ecology. Univ. Park Press, Baltimore. 227pp.

YARON, Z. 1964. Notes on the ecology and entomostracan fauna of temporary rainpools in Israel. Hydrobiologia 24: 489-513.

YOUNG, F. N. 1960. The water beetles of temporary ponds in southern Indiana. Proc. Indiana Acad. Sci. 69: 154-164.

YOUNG, F. N. & J. R. ZIMMERMAN. 1956. Variations in temperature in small aquatic situations. Ecology 37: 609-611.

ZEDLER, P. H. 1981. Microdistribution of vernal pool plants of Kearny Mesa, San Diego County. pp. 185-197. *In*: Vernal pools and intermittent streams (S. Jain and P. Moyle, eds.), Institute of Ecology, Univ. California, Davis, Pub. No. 28.

ZETTEL, J. 1984. The significance of temperature and barometric pressure changes for the snow surface activity of *Isotoma hiemalis* (Collembola). Experienta 40: 1369-1372.

# INDEX